MATHEMATICAL PROGRAMMING STUDY 17

A PUBLICATION OF THE MATHEMATICAL PROGRAMMING SOCIETY

Nondifferential and Variational Techniques in Optimization

Proceedings of the Workshop on Numerical Techniques for Systems Engineering Problems, Part 2

Edited by D.C. SORENSEN and R.J.-B. WETS

D.P. Bersekas
R.W. Cottle
R. Fletcher
E.M. Gafni
J.-B. Hiriart-Urruty
I. Kaneko
P. Lötstedt
R. Mifflin
J.S. Pang
R.T. Rockafellar
G. Strang
R. Temam
R.S. Womersley

1982

NORTH-HOLLAND PUBLISHING COMPANY – AMSTERDAM · NEW YORK · OXFORD

This book is also available in journal format on subscription.

ISBN: 0 444 86392 3

Published by:

NORTH-HOLLAND PUBLISHING COMPANY
AMSTERDAM · NEW YORK · OXFORD

Sole distributors for the U.S.A. and Canada:

Elsevier Science Publishing Company, Inc.
52 Vanderbilt Avenue
New York, N.Y. 10017

Library of Congress Cataloging in Publication Data

Workshop on Numerical Techniques for Systems
Engineering Problems (1980 : Lexington, Ky.)
Nondifferential and variational techniques in
optimization.

(Mathematical programming study ; 17)
"A Publication of the Mathematical Programming
Society."
1. Mathematical optimization--Congresses.
2. Systems engineering--Congresses. I. Sorensen,
D. C. (Danny C.) II. Wets, Roger J.-B. III. Title.
IV. Series.
QA402.5.W67 1980 519.4 82-3512
ISBN 0-444-86392-3 (Elsevier North-Holland)
AACR2

PRINTED IN THE NETHERLANDS

NONDIFFERENTIAL AND VARIATIONAL
TECHNIQUES IN OPTIMIZATION

MATHEMATICAL PROGRAMMING STUDIES

NORTH-HOLLAND PUBLISHING COMPANY – AMSTERDAM · NEW YORK · OXFORD

PREFACE

Control and filtering problems are intrinsically optimization problems. Either one seeks to control a dynamical systems so as to *minimize* (or maximize) a given performance criterion, or in filtering and related problems, one seeks to find the *best* estimate, of the state at time t of a dynamical system perturbed by a stochastic noise process, on the basis of information collected up to time t. So, at least in theory, one could rely on optimization techniques, in particular nonlinear (and linear) programming algorithms to compute their solutions. In practice, however, there are significant hurdles to overcome. Although control and, in particular, filtering-type problems are highly structured optimization problems, their intrinsic size makes it impractical to rely on standard nonlinear programming procedures. In addition, because of its nature the problem might demand a solution in "feedback" form by which we mean that the control to be used at time t or the state-estimate at time t can be viewed (actually computed) as a "simple" adjustment to the control or state-estimate at time t-Δt.

We illustrate this situation in the framework of a (linear) filtering problem in discrete time. The state dynamics of a system are described by the (vector) difference equations for $t = 1, \dots, T$

$$x_{t+1}(\omega) = A_t x_t(\omega) + B_t w_t(\omega)$$

with initial condition

$$x_1(\omega) = w_0(\omega)$$

where the $\{w_t, t = 0, \dots, T\}$ are independent random vectors with Gaussian distribution. Instead of the actual state of the system x_t, one observes a vector y_t given by the system of equations

$$y_t(\omega) = C_t x_t(\omega) + D_t w_t'(\omega).$$

Let $\hat{x}_t(\cdot)$ denote the best estimate of the system on the basis of the observations $y_1, \dots, y_{t-1}$, best meaning here a vector-valued function (estimator) $\omega \mapsto \hat{x}_t(\omega)$ whose value for (almost) all ω depends only on the values of $y_1(\omega), \dots, y_{t-1}(\omega)$, and which minimizes $\mathbf{E}[\|\hat{x}_t(\omega) - x_t(\omega)\|^2]$ or some similar criterion function. The filtering problem is not 'solved' by finding for each t, the best estimate of x_t but finding a recursive relation that allows us to compute $\hat{x}_t(\cdot)$ as a function of $\hat{x}_{t-1}(\cdot)$ and the newly acquired information y_{t-1}. It turns out that these optimal estimators can be found if a priori we determine the so-called (filtering) gain matrices. These can be found by solving a certain matrix Ricatti equation. The basic relations, and the resulting solution method, were discovered by R. Kalman in the early 60's [1]. But his method has poor numerical properties, in particular it lacks numerical stability. It is only relatively recently that numerical methods have been devised that are satisfactory (and elegant and efficient) from a numerical viewpoint. First it was shown that the problem can be formulated as a weighted least square problem, an optimization problem of a very special type. The special structure of such a problem was exploited by C. Paige and M. Saunders [2] to reduce the original problem to an almost classical problem in numerical linear algebra, for which stable and reliable (computer-implemented) solution techniques are actually available.

This volume and its companion, that together constitute the Proceedings of the Workshop on Numerical Techniques for Systems Engineering Problems held in Lexington, Kentucky (June 1980), are dedicated to the premise that modern techniques of linear algebra and non-classical optimization could be exploited to devise computational schemes for systems engineering problems that have all the desirable numerical qualifications. In Part 1 of this Proceedings we record actual advances made in this direction as well as some recent theoretical developments in systems theory that appear to be intimately related. The variational and optimization techniques that appear to promise further development are reviewed in this volume.

The first part of this volume deals with nondifferentiable optimization theory and algorithms. In "Characterizations of the plenary hull of the generalized Jacobian matrix", Jean-Baptiste Hiriart-Urruty studies the properties of the generalized Jacobian matrix of a nonlinear transformation. The relationships between various possible definitions are fully explored. The papers of R.T. Rockafellar and Robert Womersley both deal with optimality conditions. Whereas Womersley derives second order necessary and sufficient conditions for problems involving piecewise smooth functions, Rockafellar essentially limits himself to first order conditions but for problems whose constraints and objective are locally Lipschitz. He does this by studying the subgradients (and subderivatives) of the marginal function $p(u) = \inf(P_u)$ where P_u results from the perturbation of the original problem by some parameters u.

Algorithmic procedures for nondifferentiable optimization problems are described by Roger Fletcher and Robert Mifflin. The use of nondifferential techniques in the design of control systems subject to singular value inequalities is illustrated by the article of D. Mayne and E. Polak that appears in Study 18 [3]. In "A model algorithm for composite nondifferentiable optimization problems", Fletcher considers the minimization of composite functions of the type $f(\cdot)+h(G(\cdot))$ where both f and G are smooth maps and h is convex. From a nonlinear optimization viewpoint this class of functions is quite general since it encompasses exact penalty functions, nonlinear min-max functions, and best approximations.

Relying on a linear approximation to G and a quadratic approximation to f, Fletcher proves that the method has a second order rate of convergence. He also shows that his algorithm converges globally if one incorporates a trust region restriction on the step size.

Mifflin's algorithm is a modification of an earlier algorithm of C. Lemarechal. This new method will solve a wide variety of optimization problems with only minimal restrictions on the allowable type of constraints or objective function.

The second part of this volume deals with the applications of variational principle, solutions methods of variational inequalities and complementarity problems. In "A problem in capillarity and plasticity", Gilbert Strang and Roger Teman examine two problems that turn out to be the same; they both involve transition from equilibrium to collapse. The analysis relies on the relation between the original problem and a dual variational problem that yields the critical value. Variational principles, in the form of linear complementarity, play the key role in the analysis of time-dependent contact problems in rigid body mechanics as done by P. Lötstedt. He also outlines a numerical procedure for friction free problems. I. Kaneko uses complementarity techniques, more specifically parametric linear complementarity and quadratic programming problems, to discuss structural engineering problems that arise in the design of elastic-plastic and the analysis of elastic-rigid structures.

Solution techniques for structured large-scale linear complementarity problems and for (finite dimensional) variational inequalities are described in the last two papers. In "On the convergence of a block successive overrelaxation method for a class of linear complementarity problems", Richard Cottle and Jon-Shi Pang are motivated by the need to solve large scale capacitated quadratic transportation problems. Their results complement those of Cottle, Golub and Sacher who had considered a successive overrelaxation in a similar setting but with more restrictive assumptions. Dimitri Bertsekas and Eli Gafni study a projection method for finding the solution of the variational inequality

$$(x - v)'\bar{T}(v) \geq 0 \quad \text{for all } x \in K,$$

where K is a polyhedron and $\bar{T} = A'TA$, A is linear and T is Lipschitz continuous. The method is shown to yield an effective algorithm for certain traffic assignment problems.

The Workshop received the financial support of the Graduate School of the University of Kentucky and was sponsored by the Mathematical Programming Society. The organization of the Workshop and the preparation of these two volumes has relied heavily upon the secretarial and administrative skills of Ms. Sandy Leachman. We wish to take this opportunity to express our gratitude to her.

References

[1] R. Kalman, "A new approach to linear filtering and prediction problems", *ASME Transactions, Journal of Basic Engineering* 82D (1960) 35–45.
[2] C. Paige and M. Saunders, "Least squares estimation of discrete linear dynamic systems using orthogonal transformations", *SIAM Journal on Numerical Analysis* 14 (1977) 180–193.
[3] D.C. Sorensen and R.J.-B. Wets, eds., *Algorithms and theory in filtering and control*, Mathematical Programming Studies, Vol. 18 (North-Holland, Amsterdam, 1982).

Danny C. Sorensen
Argonne National Laboratory

Roger J-B. Wets
University of Kentucky

CONTENTS

Mathematical Programming Study 17 (1982) 1–12.
North-Holland Publishing Company

CHARACTERIZATIONS OF THE PLENARY HULL OF THE GENERALIZED JACOBIAN MATRIX

J.-B. HIRIART-URRUTY*

Département de Mathematiques Appliquées, Université de Clermont-Ferrand II, 63170 Aubiere, France

Received 22 July 1980
Revised manuscript received 26 November 1980

The generalized Jacobian matrix of a locally Lipschitz mapping is a set of matrices which plays a role similar to that of the Jacobian matrix for differentiable mappings. In this paper we give various characterizations of the plenary hull of the generalized Jacobian matrix by considering its support bifunction, a concept which plays a role similar to that of the support function for generalized gradients of real-valued functions.

Key words: Locally Lipshitz Mappings, Generalized Derivatives, Plenary Hull.

0. Purpose and scope

There have been a number of approaches recently towards developing a set-valued derivative for locally Lipschitz mappings which generalizes the usual notion of derivative in such a way the main theorems of analysis also extend. For nonsmooth real-valued functions, Clarke's *generalized gradients* [2] have been proven to be a useful concept from the optimization viewpoint as well as from the analysis viewpoint. A major tool to deal with generalized gradients was the analytic form of their support functions. Given a real-valued f Lipschitz around x_0, the primary construction of the generalized gradient $\partial f(x_0)$ of f at x_0 is directly related to ordinary gradients of f at those points around x_0 where f is differentiable. One of the key results of Clarke's work was the description of the support function of $\partial f(x_0)$ in terms of limits of quotients associated with f. This support function $u \mapsto f^0(x_0; u)$ which is the so-called generalized directional derivative of f at x_0 has been a major tool to derive calculus rules and also to pave the way to various generalizations for nonlocally Lipshitz functions.

For vector-valued mappings, things are less pleasant. The different attempts have been either *description of properties* which would secure the extension of usual theorems in analysis [4, 5, 23] or *direct constructions* which heavily depend on the use of derivatives in the usual sense [3, 15]. Whatever the description or construction of the set-valued derivatives, all of them act like derivatives, i.e., they share many of the properties of derivatives and actually are reduced to the derivative in certain situations. Clarke's *generalized Jacobian matrix* [3] was

* At present: Université Paul Sabatier (Toulouse III), 118, route de Narbonne, 31062 Toulouse Cedex, France.

defined for locally Lipschitz mappings following a process identical to that carried out in the real-valued case. The desire to make the generalized Jacobian matrix blind to sets of measure zero led Pourciau [15] to slightly alter Clarke's original definition. Actually there is no difference between Clarke's and Pourciau's objects as far as images of any vector are considered. We will discuss this point thoroughly in Section 2. Anyway, in spite of its rough definition, the generalized Jacobian matrix has been proved to be useful in deriving inverse function theorems, implicit function theorems, interior mapping theorems, etc... There have been further (and more general) extensions of the notion of derivative by Warga [23] who develops the concept of *derivative container* and by Halkin [4, 5] who deals with what he calls *screens*. Those objects are defined through derivatives of special approximations of the mapping around the considered point. For that, it is not necessary to assume the mappings considered locally Lipschitz. Nevertheless, we retain Clarke's generalized Jacobian matrix for its elementary definition.

To make the concept of generalized Jacobian matrix easier to handle, it is of importance to produce an analytic expression of its support function. We do not claim to have fully answered this question in the present paper. However we put in perspective the role played by a certain *support bifunction* whose role looks like that of a support function. Given a mapping F assumed Lipschitz around x_0, the generalized Jacobian matrix of F at x_0 is a convex compact set $\mathcal{J}F(x_0)$ of matrices whose support bifunction is defined as

$$(u, v) \mapsto \max_{M \in \mathcal{J}F(x_0)} \langle Mu, v \rangle.$$

This bifunction, denoted by $F^0(x_0; u, v)$, can easily be described in terms of limits of quotients associated with F. The drawback is that $F^0(x_0; \cdot, \cdot)$ is not attached with $\mathcal{J}F(x_0)$ but with the set of those matrices M for which $Mu \in \mathcal{J}F(x_0)u$ for all u. That is the reason why we are naturally led to considering the *plenary hull* of $\mathcal{J}F(x_0)$, a device due to Sweetser [19, 20] and which intends to take into account those matrices which verify the above mentioned property. The bifunction $(u, v) \mapsto F^0(x_0; u, v)$ can, moreover, be viewed under different angles: as a function of u and as a function of v. This way of doing yields divers interpretations of the plenary hull of $\mathcal{J}F(x_0)$.

In our paper, we exclusively deal with a *finite-dimensional setting*. In the very recent years, there has been a lot of works devoted to the problem of defining generalized derivatives for nonsmooth mappings from a Banach space into another. All these approaches, carried out in terms of sets of linear continuous mappings or set-valued mappings of a special structure, have as a common underlying canvas what is known for locally Lipschitz mappings. In Section 3, we briefly outline those definitions which are most in line with the development of this paper. Connections with related works are also mentioned.

1. Preliminary definitions

In this paper, $\mathbb{R}^m$ will be the vector space of real m-tuples with the usual inner product denoted by $\langle \cdot , \cdot \rangle$. The components of a function F taking values in $\mathbb{R}^q$ are indicated by $f_1, \ldots, f_q$ and $F(x)$ is assumed to be represented by the column vector $(f_1(x), \ldots, f_q(x))^{\mathrm{T}}$. When F is differentiable at $x \in \mathbb{R}^p$, the matrix representation of $F'(x)$ with respect to the canonical bases of $\mathbb{R}^p$ and $\mathbb{R}^q$ is given by the Jacobian matrix we denote by $JF(x)$. In the special case when f is real-valued, $f'(x)$ is represented by a row vector and the column vector $f'(x)^{\mathrm{T}}$, denoted by $\nabla f(x)$, is the gradient of f at x.

Let $\mathcal{O}$ be a nonempty open subset of $\mathbb{R}^p$, let f be a real-valued locally Lipschitz function and $x_0 \in \mathcal{O}$; Clarke [2] defined the *generalized gradient* of f at x_0, denoted $\partial f(x_0)$, by taking the convex hull of $\underset{\to}{\nabla} f(x_0)$, where $\underset{\to}{\nabla} f(x_0)$ stands for the set of limits of the form $\lim_{n\to\infty} \nabla f(x_n)$, where f is differentiable at x_n and x_n converges to x_0. The nonempty compact convex set obtained in such a way is completely characterized by its support function

$$u \in \mathbb{R}^p \mapsto f^0(x_0; u) = \max_{x^* \in \partial f(x_0)} \langle x^*, u \rangle, \tag{1.1}$$

and a major result in [2] was that $f^0(x_0; u)$ could be expressed in terms of quotient limits, namely

$$f^0(x_0; u) = \limsup_{\substack{x \to x_0 \\ \lambda \to 0^+}} \frac{f(x + \lambda u) - f(x)}{\lambda}.$$

Let now F be locally Lipschitz on $\mathcal{O} \subset \mathbb{R}^p$ and taking values in $\mathbb{R}^q$. One could be tempted to define the generalized derivative of $F = (f_1, \ldots, f_q)^{\mathrm{T}}$ at x_0 by simply considering

$$[\partial f_1(x_0), \ldots, \partial f_q(x_0)]^{\mathrm{T}}, \tag{1.2}$$

where the above set consists of matrices whose ith row belongs to $\partial f_i(x_0)$. Actually, the resulting set is too large and one easily realizes that the possible 'interdependance' of component functions f_i is lost in such a way of doing. A sharper mathematical tool is what Clarke [3] called the *generalized Jacobian (matrix)* defined in the following way:

Definition 1.1. The generalized Jacobian matrix of F at $x_0 \in \mathcal{O}$, denoted by $\mathcal{J}F(x_0)$, is the convex hull of $\underset{\to}{J}F(x_0)$, where

$$\underset{\to}{J}F(x_0) = \{\underset{\to}{J} \mid \exists\, x_n \to x_0 \text{ in dom } F',\ \underset{\to}{J} = \lim_{n\to\infty} JF(x_n)\}. \tag{1.3}$$

In this definition, dom F' denotes the subset of full measure of $\mathcal{O}$ where F is differentiable.

In doing so, $\mathcal{J}F(x_0)$ is a *nonempty compact convex* subset of the space $\mathcal{M}(q,p)$ of (q,p) matrices, which is reduced to $\{JF(x_0)\}$ wherever F is strictly differentiable at x_0. For example, if $F=(f,\dots,f)^{\mathrm{T}}$ consists of q copies of the same real-valued f, one has that

$$\mathcal{J}F(x_0)=\{[x^*,\dots,x^*]^{\mathrm{T}}\mid x^*\in\partial f(x_0)\}.$$

The basic properties of the set-valued mapping $x\rightrightarrows\mathcal{J}F(x)$ are displayed in [3]. For examples of its use in nonsmooth optimization, one could refer to the author's recent papers [9, 10].

2. The plenary hull of $\mathcal{J}F(x_0)$

To make the concept of generalized Jacobian matrix more workable, it would be of great interest to derive the support function of $\mathcal{J}F(x_0)$ in the linear space $\mathcal{M}(q,p)$. For that purposes, let us denote by $\langle\langle\cdot,\cdot\rangle\rangle$ the inner product on $\mathcal{M}(q,p)$ defined by $\langle\langle M,U\rangle\rangle=$ trace of $M\circ U^{\mathrm{T}}$; it comes from Definition 1.1 that for all $U\in\mathcal{M}(q,p)$:

$$\max_{M\in\mathcal{J}F(x_0)}\langle\langle M,U\rangle\rangle=\limsup_{\substack{x\to x_0\\ x\in\operatorname{dom}F'}}\langle\langle JF(x),U\rangle\rangle. \tag{2.1}$$

By describing U in terms of $u_i\in\mathbb{R}^p$ as $[u_1,\dots,u_q]^{\mathrm{T}}$, the above expression can be rewritten as:

$$\max_{M\in\mathcal{J}F(x_0)}\langle\langle M,U\rangle\rangle=\limsup_{\substack{x\to x_0\\ x\in\operatorname{dom}F'}}\sum_{i=1}^{q}\langle\nabla f_i(x),u_i\rangle. \tag{2.2}$$

The next step would be now to translate the right-hand side of (2.2) in terms of difference quotients of some kind involving the f_i. This operation turns out to be difficult for general U whenever $\min(p,q)>1$. That is the reason why we turn our attention to those $U\in\mathcal{M}(q,p)$ of the form $u\otimes v:x\mapsto\langle u,x\rangle v$, where $u\in\mathbb{R}^p$ and $v\in\mathbb{R}^q$. In such a way, $\langle\langle M,u\otimes v\rangle\rangle$ reduces to $\langle Mu,v\rangle$ and (2.1) can be rephrased as:

$$\max_{M\in\mathcal{J}F(x_0)}\langle Mu,v\rangle=\limsup_{\substack{x\to x_0\\ x\in\operatorname{dom}F'}}\langle JF(x)u,v\rangle. \tag{2.3}$$

Now, we can proceed in two ways: either translate the right-hand side of (2.3) by using the perturbation techniques developed in [2, 3], or use existing results on chain rules so that the left-hand side of (2.3) appears as the generalized gradient of a particular real-valued function. We operate in the latter way. Given $v\in\mathbb{R}^q$, the generalized gradient of $F_v:x\mapsto\langle F(x),v\rangle$ at x_0 can be *exactly* described as:

$$\partial F_v(x_0)=\mathcal{J}^{\mathrm{T}}F(x_0)v \qquad [7,8].$$

Therefore, for all $u \in \mathbb{R}^p$, we have that:

$$\max_{M \in \mathcal{J}F(x_0)} \langle u, M^{\mathrm{T}} v \rangle = F_v^0(x_0; u). \tag{2.4}$$

Hence, since we know the analytic expression of $F_v^0(x_0; \cdot)$, we have the following.

Theorem 2.1. *Let $u \in \mathbb{R}^p$ and $v = (v_1, \dots, v_q) \in \mathbb{R}^q$. Then*

$$\max_{M \in \mathcal{J}F(x_0)} \langle Mu, v \rangle = \limsup_{\substack{x \to x_0 \\ \lambda \to 0^+}} \sum_{i=1}^{q} \frac{v_i[f_i(x + \lambda u) - f_i(x)]}{\lambda}. \tag{2.5}$$

This result does not fully answer our question of determining the support function of $\mathcal{J}F(x_0)$. Of course, all $U \in \mathcal{M}(q, p)$ can be expressed as

$$U = \sum_{k=1}^{r} u_k \otimes v_k,$$

where $u_k \in \mathbb{R}^p$, $v_k \in \mathbb{R}^q$ and $r = \min(p, q)$. The general desired result would be an expression of

$$\max_{M \in \mathcal{J}F(x_0)} \sum_{k=1}^{r} \langle Mu_k, v_k \rangle, \tag{2.6}$$

while we only are able to produce $\max_{M \in \mathcal{J}F(x_0)} \langle Mu_k, v_k \rangle$ for each k.

Nevertheless, the function $(u, v) \mapsto F^0(x_0; u, v) = F_v^0(x_0; u)$ has some special features we are going to describe now. Before going further, we may particularize the cases where p or q equals 1. If $q = 1$ (i.e., $F = f$ is real-valued), we recognize known results in such a situation since:

$$F^0(x_0; u, v) = \begin{cases} vf^0(x_0; u), & \text{if } v \geq 0, \\ -vf^0(x_0; -u), & \text{if } v \leq 0. \end{cases} \tag{2.7}$$

When $p = 1$, $F = \boldsymbol{x}$ is what can be called a 'locally Lipschitz curve' in $\mathbb{R}^q$,

$$\boldsymbol{x} :]\alpha, \beta[\to \mathbb{R}^q$$
$$t \mapsto \boldsymbol{x}(t).$$

In such a case, the generalized derivative $\partial \boldsymbol{x}(t_0)$ of $\boldsymbol{x}$ at $t_0 \in]\alpha, \beta[$ is completely characterized since

$$F^0(t_0; u, v) = \max_{x^* \in \partial \boldsymbol{x}(t_0)} \langle x^*, uv \rangle$$

for all $u \in \mathbb{R}$ and $v \in \mathbb{R}^q$, so that Theorem 2.1 can be rephrased as:

$$x^* \in \partial \boldsymbol{x}(t_0) \Leftrightarrow \begin{cases} \langle x^*, v \rangle \leq \limsup\limits_{\substack{t \to t_0 \\ \lambda \to 0^+}} \dfrac{\langle \boldsymbol{x}(t + \lambda) - \boldsymbol{x}(t), v \rangle}{\lambda}. \\ \text{for all } v \in \mathbb{R}^q. \end{cases} \tag{2.8}$$

The first salient feature of the formula (2.5) in Theorem 2.1 is that we have a full characterization of the *images* of any vector u by $\mathcal{J}F(x_0)$ in terms of difference quotients involving component functions f_i. Actually, the function $v \mapsto F^0(x_0; u, v)$ is nothing than the support function of $\mathcal{J}F(x_0)u$. At this stage of the discussion, there is a point which is worth mentioning and which deals with the (possible) alteration of the generalized Jacobian matrix when, in the definition of $\mathcal{J}F(x_0)$, we impose that x_n lies in the complement of a given set of null measure.

Proposition 2.2. *Let $\Lambda \subset \operatorname{dom} F'$ be such that its complementary set in $\mathcal{O}$ is of null measure, and let $\mathcal{J}_\Lambda F(x_0)$ be defined as in* (1.3) *except that in this definition the points x_n are constrained in Λ. Then*

$$\max_{M \in \mathcal{J}_\Lambda F(x_0)} \langle Mu, v\rangle = F^0(x_0; u, v) \tag{2.9}$$

for all $u \in \mathbb{R}^p$ and $v \in \mathbb{R}^q$, so that

$$\mathcal{J}_\Lambda F(x_0)u = \mathcal{J}F(x_0)u \tag{2.10}$$

for all $u \in \mathbb{R}^p$.

Proof. Due to the definitions themselves, we have that

$$\partial_\Lambda F_v(x_0) = \mathcal{J}_\Lambda^{\mathrm{T}} F(x_0)v.$$

It is known that the generalized gradient of real-valued functions is blind to sets of measure zero [2, Proposition 1.11]. Hence $\partial_\Lambda F_v(x_0)$ equals $\partial F_v(x_0)$ and the announced results are proved.

The result is of importance since it shows that as far as the *images* are concerned, one can ignore 'thin' subsets in the construction of $\mathcal{J}F(x_0)$. That does not imply a priori that $\mathcal{J}_\Lambda F(x_0)$ equals $\mathcal{J}F(x_0)$, except at least in two particular cases, namely when p or q equals 1. Indeed, in these cases, the knowledge of the images $\{\mathcal{J}F(x_0)u \mid u \in \mathbb{R}^p\}$ is tantamount to the full knowledge of $\mathcal{J}F(x_0)$. So, except in the above mentioned cases, the equality (2.10) does not allow us to claim that $JF(x_0)$ itself is blind to sets of null measure. The desire to make the generalized derivative blind to sets of null measure led Pourciau to alter Clarke's original definition by considering the Lebesgue set $\operatorname{Leb} F'$ of F', instead of $\operatorname{dom} F'$, in the definition of $\mathcal{J}F(x_0)$ [15]. In doing so, Pourciau derived a concept of generalized Jacobian matrix, denoted by $\mathcal{J}^{\mathrm{P}}F(x_0)$, enjoying many properties of Clarke's generalized Jacobian matrix. The Lebesgue point restriction, although technical and difficult to handle, seemed worthwhile since it made $\mathcal{J}^{\mathrm{P}}F(x_0)$ invariant if one ignores sets of null measure [15, Proposition 4.2]. But, since F' is locally in $L^\infty(\mathcal{O}, \mathbb{R}^q)$, almost every x in $\operatorname{dom} F'$ belongs to $\operatorname{Leb} F'$. With this in mind, Proposition 2.2 induces that

$$\mathcal{J}^{\mathrm{P}}F(x_0)u = \mathcal{J}F(x_0)u$$

for all $u \in \mathbb{R}^p$. In short, *there is no gap between Clarke's and Pourciau's definitions as far as images are concerned.* As for the set of matrices themselves, $\mathcal{J}^P F(x_0)$ is a priori included in $\mathcal{J}F(x_0)$; however a gap (if any!) might exist only in some pathological cases [12].

In Theorem 2.1, we secured that all of $M \in \mathcal{J}F(x_0)$ do satisfy

$$\langle Mu, v\rangle \leq F^0(x_0; u, v)$$

for all $(u, v) \in \mathbb{R}^p \times \mathbb{R}^q$. Conversely, what are the $M \in \mathcal{M}(q, p)$ which satisfy the above inequality for all (u, v)? Since $\mathcal{J}F(x_0)u$ is closed and convex for all u, the question can be formulated in an equivalent way as: What are the $M \in \mathcal{M}(q, p)$ such that

$$Mu \in \mathcal{J}F(x_0)u$$

for all $u \in \mathbb{R}^p$? Although $\mathcal{J}F(x_0)$ is convex and compact, one generally cannot separate an M_0 from $\mathcal{J}F(x_0)$ by using only linear mappings (on $\mathcal{M}(q, p)$) of the form $u \otimes v$, $u \in \mathbb{R}^p$ and $v \in \mathbb{R}^q$. This state of affairs led Halkin and Sweetser [19, Section 3] to introduce the following definition: a subset $\mathcal{A} \subset \mathcal{M}(q, p)$ is *plenary* if and only if it includes every $M \in \mathcal{M}(q, p)$ satisfying $Mu \in \mathcal{A}u$ for all $u \in \mathbb{R}^p$. In Rubinov's terminology, such sets are called *solid* [18]. Since the intersection of plenary sets is plenary, Halkin and Sweetser defined the *plenary hull of* $\mathcal{A}$, denoted plen $\mathcal{A}$, as the smallest plenary set containing $\mathcal{A}$. The concept of plenarity in an infinite-dimensional setting and the properties of plenary sets from both algebraic and topological viewpoints are studied in Sweetser's thesis [20, Chapter IV] where, moreover, are displayed some funny examples. In fact, these definitions were introduced for allowing to transcribe in terms of the linear operators themselves the information contained in the collection of images. When $q = 1$, the convexity of $\partial f(x_0)$ suffices to ensure that it is plenary (plen $\partial f(x_0) = \partial f(x_0)$). Otherwise, namely when $\min(p, q) > 1$, plen $\mathcal{J}F(x_0)$ is a *convex compact* (*plenary*) set of matrices containing $\mathcal{J}F(x_0)$.

Since $\mathcal{J}F(x_0)u = [\text{plen}\, \mathcal{J}F(x_0)]u$ for all $u \in \mathbb{R}^p$, the following is a reformulation of Theorem 2.1.

Proposition 2.3. *Let* $u \in \mathbb{R}^p$ *and* $v \in \mathbb{R}^q$. *Then*

$$\max_{M \in \text{plen}\, \mathcal{J}F(x_0)} \langle Mu, v\rangle = F^0(x_0; u, v). \tag{2.11}$$

In another setting, $M \in \text{plen}\, \mathcal{J}F(x_0)$ *if and only if* $\langle Mu, v\rangle \leq F^0(x_0; u, v)$ *for all* $(u, v) \in \mathbb{R}^p \times \mathbb{R}^q$.

Formulated in such a manner, the second statement of the above result appears in a more or less hidden form in Sweetser's paper [19, Section 4]. Sweetser's approach consisted in considering first what he called a *shield* of F at x_0. $\mathfrak{A} \subset \mathcal{M}(q, p)$ is a shield for F at x_0 if and only if, for any $\epsilon > 0$, there is a $\eta > 0$

such that, for any distinct x_1 and x_2 in the neighborhood $B(x_0, \eta)$ of x_0, there is $A \in \mathfrak{A}$ such that

$$\|F(x_1) - F(x_2) - A(x_1 - x_2)\| \leq \epsilon \|x_1 - x_2\|.$$

What Sweetser showed is that plen $\mathcal{J}F(x_0)$ is the *unique minimal convex, closed, plenary* shield of F at x_0 [19, Section 5].

In terms of plenary hull, Proposition 2.2 can be reformulated as follows:

Proposition 2.4. *Let Λ and $\mathcal{J}_\Lambda F(x_0)$ be as in Proposition 2.2. Then*

$$\text{plen}\, \mathcal{J}_\Lambda F(x_0) = \text{plen}\, \mathcal{J}F(x_0). \tag{2.12}$$

So, plen $\mathcal{J}F(x_0)$ is blind to sets of measure zero and, in particular, Clarke's $\mathcal{J}F(x_0)$ and Pourciau's $\mathcal{J}^{\text{P}}F(x_0)$ yield the same plenary hull. Even if one does not have an analytic description of plen $\mathcal{J}F(x_0)$ other than its definition, no irregularity is introduced by taking the plenary hull, as shown by Sweetser [19, Corollary 5.2]. Indeed, if all $M \in \mathcal{J}F(x_0)$ are *onto*, so are all $M \in \text{plen}\, \mathcal{J}F(x_0)$. This key result explains the different twists given to certain assumptions in nonsmooth analysis [12]. For example, the inverse function theorem for locally Lipschitz mappings [3] remains true if one imposes the maximality of rank only of all $M \in \mathcal{J}_\Lambda F(x_0)$ rather than those of $\mathcal{J}F(x_0)$ [3, Remark 5]; the reason is that the intrinsic assumption turns out to be the maximality of rank of all $M \in \text{plen}\, \mathcal{J}F(x_0)$.

Now, let us go back to formula (2.5) in Theorem 2.1. Until now, we have looked at $F^0(x_0; u, v)$ mainly as a function of v for a fixed u, i.e., as the support function of $\mathcal{J}F(x_0)u$. For a fixed v, we already have noticed that $u \mapsto F^0(x_0; u, v)$ is the support function of $\mathcal{J}^{\text{T}}F(x_0)v$. What we intend developing in the next interpretation is that one can be led to the concept of plenarity by looking at the problem under a different angle. Since for all $v \in \mathbb{R}^q$, the real-valued function $F_v : x \mapsto \langle F(x), v\rangle$ is locally Lipschitz, it is quite natural to pose the problem of determining the compact convex sets $\mathcal{K}(F; x_0)$ of $\mathcal{M}(p, q)$ satisfying

$$\partial F_v(x_0) = \mathcal{K}(F; x_0)v \tag{2.13}$$

for all $v \in \mathbb{R}^q$. Such compact convex sets do exist since $\mathcal{J}_\Lambda^{\text{T}}F(x_0)$ satisfy the required property for any Λ as in Proposition 2.2. Now, a question which arises from this formulation is the following: what is the *maximal* convex compact set $\mathcal{K}(F; x_0)$ satisfying the above requirement?

Proposition 2.5. $\mathcal{K}(F; x_0) = [\text{plen}\, \mathcal{J}F(x_0)]^{\text{T}} = \text{plen}\, \mathcal{J}^{\text{T}}F(x_0)$ *is the maximal convex compact set of $\mathcal{M}(p, q)$ for which $\partial F_v(x_0) = \mathcal{K}(F; x_0)v$ for all $v \in \mathbb{R}^q$.*

The proof is just a matter of giving another twist to the expression of $F^0(x_0; u, v)$ in Proposition 2.3.

To summarize, let's say that plen $\mathcal{J}F(x_0)$ is the maximal convex compact (plenary) set of matrices satisfying $[\text{plen}\,\mathcal{J}F(x_0)]u = \mathcal{J}F(x_0)u$ for all $u \in \mathbb{R}^p$. When $F = (f_1, \dots, f_q)^T$ we have that:

$$\mathcal{J}F(x_0) \subset \text{plen}\,\mathcal{J}F(x_0) \subset [\partial f_1(x_0), \dots, \partial f_q(x_0)]^T. \tag{2.14}$$

The set $[\partial f_1(x_0), \dots, \partial f_q(x_0)]^T$ is obviously convex, compact *and plenary.* It actually yields the same image set as $\mathcal{J}F(x_0)$ does when the considered vectors u are the elements e_i of the canonical basis of $\mathbb{R}^p$. In other words,

$$\{x_i^*, [x_1^*, \dots, x_i^*, \dots, x_q^*]^T \in \mathcal{J}F(x_0)\} = \partial f_i(x_0) \qquad [7,8].$$

Let us illustrate the double inclusion (2.14) with an example derived from [19, Example 6.2].

Let $F = (f_1, f_2)^T : \mathbb{R}^2 \to \mathbb{R}^2$ defined as follows:

$$F(x_1, x_2) = \begin{cases} (x_1, -x_2)^T, & \text{if } x_1 \geq 0, \\ (x_1, -2x_1 - x_2)^T, & \text{if } x_1 \leq 0 \text{ and } x_2 \geq -x_1, \\ (-2x_1 - 2x_2, x_2)^T, & \text{if } x_1 \leq 0 \text{ and } 0 \leq x_2 \leq -x_1, \\ (-x_1 + 2x_2, x_2)^T, & \text{if } x_1 \leq 0 \text{ and } 0 \geq x_2 \geq x_1, \\ (x_1, 2x_1 - x_2)^T, & \text{if } x_1 \leq 0 \text{ and } x_2 \leq x_1. \end{cases}$$

F is a piecewise linear function and the generalized Jacobian matrix at 0 is given by

$$\mathcal{J}F(0) = \text{co}\left\{\begin{bmatrix} 1 & 0 \\ -2 & -1 \end{bmatrix}, \begin{bmatrix} -1 & -2 \\ 0 & 1 \end{bmatrix}, \begin{bmatrix} -1 & 2 \\ 0 & 1 \end{bmatrix}, \begin{bmatrix} 1 & 0 \\ 2 & -1 \end{bmatrix}\right\}.$$

Since $u \in \mathcal{J}F(0)u$ for all $u \in \mathbb{R}^2$, the identity matrix is in plen $\mathcal{J}F(0)$ [19, p. 560] while it is not in $\mathcal{J}F(0)$.

The same example serves as an illustration of the second inclusion in (2.14). Indeed,

$$\partial f_1(0) = \text{co}\left\{\begin{pmatrix} 1 \\ 0 \end{pmatrix}, \begin{pmatrix} -1 \\ 2 \end{pmatrix}, \begin{pmatrix} -1 \\ -2 \end{pmatrix}\right\}, \qquad \partial f_2(0) = \text{co}\left\{\begin{pmatrix} 0 \\ 1 \end{pmatrix}, \begin{pmatrix} 2 \\ -1 \end{pmatrix}, \begin{pmatrix} -2 \\ -1 \end{pmatrix}\right\}.$$

As for example, $\left[\begin{smallmatrix} 1 & 0 \\ 1 & 0 \end{smallmatrix}\right]$ belongs to $[\partial f_1(0), \partial f_2(0)]^T$ but is not an element of $\mathcal{J}F(0)$.

As noticed earlier, the functions $F^0(x_0; u, \cdot)$ and $F^0(x_0; \cdot, v)$ are, for all $(u, v) \in \mathbb{R}^p \times \mathbb{R}^q$, support functions of convex sets. To a certain extent, the function $(u, v) \mapsto F^0(x_0; u, v)$ is to $\mathcal{J}F(x_0)$ (and therefore to plen $\mathcal{J}F(x_0)$) what a support function is to a set of vectors. In order to avoid some possible confusion with the support function of $\mathcal{J}F(x_0)$, we set the following:

Definition 2.6. Given $\mathcal{A}$ a convex compact subset of $\mathcal{M}(q, p)$, the function $(u, v) \mapsto \max_{M \in \mathcal{A}} \langle Mu, v \rangle$ is called the support bifunction of $\mathcal{A}$.

In the present context, the support bifunction of $\mathcal{J}F(x_0)$ (and therefore of plen $\mathcal{J}F(x_0)$) is $(u, v) \mapsto F^0(x_0; u, v)$. The concept of support bifunction turns out to be useful for deriving in a clear-cut manner various results in nonsmooth analysis like: mean value theorems, general chain rules, implicit function theorems ... We hope to present them elsewhere [12].

To end this section, we pose a question whose full answer is unknown to us. Given a closed convex set C in $\mathbb{R}^p$, the projection operator $P_C:\mathbb{R}^p \mapsto \mathbb{R}^p$ is known to be Lipschitz. What is the generalized Jacobian matrix of P_C? What is its signification from the geometrical viewpoint?

3. Related works

Defining a concept of generalized derivative for locally Lipschitz mappings from a Banach space into another has been a main concern in the recent years. Let us briefly sketch those approaches which are most in line with the present report. The common framework dealt with is:

$$F: X \to Y,$$

where X and Y are real Banach spaces and F a locally Lipschitz mapping. Thibault in his thesis [22] mainly dealt with *ordered* spaces Y and, by generalizing in an appropriate manner the 'lim sup' operation involved in the definition of $f^0(x_0; u)$ for a real-valued function f, he was able to define a *generalized subdifferential* for the so-called class of *compactly Lipschitz* mappings [21]. This object denoted here by $\partial^{\otimes}F(x_0)$, turns out to be $\mathbf{X}_{i=1}^{q}\, \partial f_i(x_0)$ when $F = (f_1, \dots, f_q)^{\mathrm{T}}$ takes values in $\mathbb{R}^q$ endowed with its natural ordering. The generalization of what has been carried out in the finite-dimensional setting was however possible when X is a *separable* Banach space and Y a *reflexive separable* Banach space [22, Chapter III]. In such a setting, a generalization of Rademacher's theorem states that F is Hadamard-differentiable at all points x of a set dom DF whose complementary set is Haar-null. The notion of Haar-null sets is a generalization of the notion of sets of Lebesgue measure zero in $\mathbb{R}^p$. So, by mimicing the process carried out in finite dimensions, it is allowable to consider the set $\underset{\rightarrow}{DF}(x_0)$ of all limits of the form $\lim_{n\to\infty} DF(x_n)$, with x_n converging to x_0 in dom DF and where the limit is taken in $L_w(X, Y)$ (i.e., $L(X, Y)$ endowed with the weak operator topology). The next step consists of defining the *generalized derivative* of F at x_0, denoted $\partial F(x_0)$, as the closed convex hull of $\underset{\rightarrow}{DF}(x_0)$. Hence $\partial F(x_0)$ is a *nonempty* compact convex set of $L_w(X, Y)$ whose characterizations are similar to those displayed in the present paper. In particular, Thibault was led to considering the plenary hull of $\partial F(x_0)$ by posing the questions of existence and characterization of the maximal closed convex set $\Gamma(F; x_0)$ of $L_w(X, Y)$ such that, for all $v^* \in Y^*$,

$$\partial(v^* \circ F)(x_0) = v^* \circ \Gamma(F; x_0).$$

The answer in the (easier) case of finite-dimensional settings was given in

Proposition 2.5. For the generalization of these different characterizations, we refer the reader to the short note [11]. Finally, it is noteworthy that there is a connection between $\partial F(x_0)$ and $\partial^{\otimes}F(x_0)$ when both can be defined. Namely, $\partial^{\otimes}F(x_0)$ is the *operative convex hull* of $\partial F(x_0)$ [22, Chapter III; 18].

A quite natural generalization of the definition of $F^0(x_0; \cdot, \cdot)$ is to set

$$\forall\, u \in X,\ \forall\, v^* \in Y^*,\ F^0(x_0; u, v^*) = \limsup_{\substack{x \to x_0 \\ \lambda \to 0^+}} \frac{\langle F(x + \lambda u) - F(x), v^* \rangle}{\lambda}.$$

We therefore are led to pose the questions of the existence and the characterization of those $M \in L(X, Y)$ which satisfy

$$\forall\, u \in X,\ \forall\, v^* \in Y^*,\quad \langle Mu, v^* \rangle \leq F^0(x_0; u, v^*).$$

These questions were answered in the setting quoted just earlier. However, the mere question of existence is hopeless for X and Y general Banach spaces. This state of affairs led Ioffe [13] to define the generalized derivative of F at x_0 in terms of *set-valued mappings* of a special kind instead of (convex) sets of linear mappings. Actually, Ioffe considers the set-valued mapping $\mathbb{D}F(x_0): X \rightrightarrows Y^*$ which assigns to u the closed convex set of Y^* whose support function is precisely $F^0(x_0; u, v^*)$. In Ioffe's terminology, this set-valued mapping is an *odd bounded fan*; for more details on Ioffe's approach and the problems posed in that respect, see [1] and the very recent paper [14] we just got from Prof. Rockafellar and which develops to many extents the note [13]. Of course, in the finite-dimensional setting which was ours in the paper, all this material is unnecessary since the fan $u \rightrightarrows \mathbb{D}F(x_0)u$ is *generated* by $\mathcal{J}F(x_0)$. Still in Ioffe's terminology, observe that what Ioffe looks for as being the *handle of the fan* $\mathbb{D}F(x_0)$ is merely the plenary hull of $\mathcal{J}F(x_0)$.

There are also some close connections between Ioffe's *strict prederivatives* [14, Chapter 2] and the concepts of *shields* such as developed by Sweetser [20, Chapter 2]. To be more concise, a shield of F over a neighborhood of x_0 is a subset $\mathfrak{A} \subset L(X, Y)$ and a strict prederivative of F at x_0 is a set-valued mapping $\mathcal{A}$ of a special type (namely a bounded fan) which both satisfy a certain approximation property of F around x_0. Actually an equicontinuous shield $\mathfrak{A}$ yields a strict prederivative $\mathcal{A}u = \{Au \mid A \in \mathfrak{A}\}$, while the set of $A \in L(X, Y)$ generating a strict prederivative is an example of shield. Moreover, the procedures of constructing shields [19, Section 2] and strict prederivatives [14, Proposition 8.5] as well as the calculus rules derived from look pretty much alike.

References

[1] J.-P. Aubin, "Ioffe's fans and generalized derivatives of vector-valued maps", in: *Convex Analysis and Optimization* (Imperial College, London, 1980) to appear in "Surveys and Reference Works in Mathematics" Series, Pitman Publishers.

[2] F.H. Clarke, "Generalized gradients and applications", *Transactions of the American Mathematical Society* 205 (1975) 247–262.
[3] F.H. Clarke, "On the inverse function theorem", *Pacific Journal of Mathematics* 64 (1976) 97–102.
[4] H. Halkin "Interior mapping theorem with set-valued derivatives", *Journal d'Analyse Mathématique* 30 (1975) 200–207.
[5] H. Halkin, "Mathematical programming without differentiability" in: D.L. Russel, ed., *Calculus of Variations and Control Theory* (Academic Press, New York, 1976) pp. 279–288.
[6] H. Halkin, "Necessary conditions for optimal control problems with differentiable or nondifferentiable data", Tech. Rept., University of California, San Diego, CA (1977).
[7] J.-B. Hiriart-Urruty, "Gradients généralisés de fonctions composées. Applications", *Note aux Comptes Rendus de l'Académie des Sciences de Paris* 285, Série A (1977) 781–784.
[8] J.-B. Hiriart-Urruty, "New concepts in nondifferentiable programming", in: J.-P. Penot, ed., *Journées d'Analyse Non Convexe, Bulletin de la Sociéte Mathématique de France,* Mémoire 60 (1979) pp. 57–85.
[9] J.-B. Hiriart-Urruty, "Refinements of necessary optimality conditions in nondifferentiable programming I", *Applied Mathematics and Optimization* 5 (1979) 63–82.
[10] J.-B. Hiriart-Urruty, "Refinements of necessary optimality conditions in nondifferentiable programming II", in: M. Guignard, ed., *Optimality and stability in mathematical programming,* Mathematical Programming Study, to appear.
[11] J.-B. Hiriart-Urruty and L. Thibault, "Existence et caractérisation de différentielles généralisées d'applications localement Lipschitziennes d'un Banach séparable dans un Banach réflexif séparable", *Note aux Comptes Rendus de l'Académie des Sciences de Paris* 290, Série A (1980) 1091–1094.
[12] J.-B. Hiriart-Urruty, "Analysis of locally Lipschitz mappings in finite dimensions", in preparation.
[13] A.D. Ioffe, "Différentielles généralisées d'applications localement Lipschitziennes d'un espace de Banach dans un autre", *Note aux Comptes Rendus de l'Académie des Sciences de Paris* 289, Série A (1979) 637–640.
[14] A.D. Ioffe, "Nonsmooth Analysis: Differential calculus of nondifferentiable mappings", *Transactions of the American Mathematical Society,* to appear.
[15] B.H. Pourciau, "Analysis and Optimization of Lipschitz continuous mappings", *Journal of Optimization Theory and Applications* 22 (1977) 311–351.
[16] B.H. Pourciau, "Univalence and degree for Lipschitz continuous maps", to appear.
[17] R.T. Rockafellar, "La théorie des sous-gradients et ses applications à l'optimisation", Presses de l'Université de Montréal, Montréal (1979).
[18] A.M. Rubinov, "Sublinear operators and operator-convex sets", *Siberian Mathematical Journal* 17 (1976) 289–295.
[19] T.H. Sweetser, "A minimal set-valued strong derivative for vector-valued Lipschitz functions", *Journal of Optimization Theory and Applications* 23 (1977) 549–562.
[20] T.H. Sweetser, "A set-valued strong derivative in infinite dimensional spaces, with applications in Hilbert spaces", Ph.D. Thesis, University of California, San Diego, CA (1979).
[21] L. Thibault, "Subdifferentials of compactly Lipschitzian vector-valued functions", Séminaire d'Analyse Convexe, exposé no. 5, Université de Montpellier, Montpellier (1978).
[22] L. Thibault, "Sur les fonctions compactement Lipschitziennes et leurs applications: programmation mathématique, contrôle optimal, espérance conditionnelle", Thèse de Doctorat ès-Sciences, Université de Montpellier, Montpellier (1980).
[23] J. Warga, "Derivative containers, inverse functions and controllability" in: D.L. Russel, ed., *Calculus of Variations and Control Theory* (Academic Press, New York, 1976) pp. 13–46.

Mathematical Programming Study 17 (1982) 13–27.
North-Holland Publishing Company

OPTIMALITY CONDITIONS FOR PIECEWISE SMOOTH FUNCTIONS*

R.S. WOMERSLEY

*Department of Mathematics, University of Dundee, Dundee, DD1 4HN, Scotland.***

Received 14 August 1980
Revised manuscript received 14 September 1981

Certain nonsmooth functions are viewed as piecewise smooth functions, which are composed of a finite number of smooth functions. Second order necessary and sufficient conditions are established for this class of nonsmooth functions. However when a piecewise smooth function cannot be expressed as the maximum of its component functions there are severe limitations in the usual first order necessary conditions, and simple examples are given to illustrate these limitations. Interpretations of the multipliers arising in the first order conditions and of the curvature information in the second order conditions are also given.

Key words: Piecewise Smooth Functions, Nonsmooth Functions, Optimality Conditions, Directional Derivative, Lagrangian Function.

1. Introduction

The aim of this paper is to view certain nonsmooth functions as piecewise smooth functions and to see if stronger optimality conditions (both necessary and sufficient) can be established for such functions. The main interest is not in producing necessary optimality conditions for general nonsmooth functions (see for example [5]), but rather in seeing how much extra information can be obtained from looking at structured classes of nonsmooth functions. Stronger optimality conditions which can thus be obtained suggest how curvature information can be used in designing minimization algorithms for certain classes of nonsmooth functions.

Second order necessary and sufficient conditions are given for the class of piecewise smooth functions. However for a piecewise smooth function which cannot be expressed as the maximum of its component functions there are severe limitations in the usual necessary conditions. Simple examples are given to illustrate these limitations. An interpretation of the multipliers arising in the first order conditions as the derivatives of the nonsmooth function with respect to its component functions is given. Also the curvature information in the

* This paper was prepared while the author was visiting the University of Kentucky, Lexington, and was partially supported by National Science Foundation Grant #ECS-7923272.

** Present address: School of Mathematics, University of New South Wales, Kensington, N.S.W., Australia 2031.

second order conditions is interpreted as the curvature of the nonsmooth function within a surface of nondifferentiability.

Any piecewise smooth function F is locally Lipschitz and so has a generalized gradient ∂F [1]. Also for the piecewise smooth functions considered it is assumed that the one-sided directional derivative $F'(x; s)$ always exists. The usual necessary condition for a local minimizer x^* of a nonsmooth function is that

$$0 \in \partial F^*. \tag{1.1}$$

However for a general piecewise smooth function, although condition (1.1) is necessary, it is 'weak' in the sense that at a point x' with $0 \in \partial F'$ there can still exist directions with $F'(x'; s) < 0$.

Let $F^0(x; s)$ denote the generalized directional derivative [1], then

$$F'(x; s) \le F^0(x; s) \quad \text{for all } x, s. \tag{1.2}$$

The weakness in the necessary condition (1.1) arises when (1.2) can be satisfied as a strict inequality. This cannot happen if the piecewise smooth function F can be expressed as the maximum of its component functions. This is precisely the case when the problem of minimizing the piecewise smooth function can be converted to a nonlinear programming problem by the addition of an extra variable. This viewpoint is not taken here as the direct approach is applicable to more general situations, leads to more natural interpretations of the results, and does not introduce any additional assumptions (for example the first order constraint qualification for this nonlinear programming problem is automatically satisfied).

The weakness in the necessary conditions (1.1) also causes problems in developing second order sufficient conditions. Even if (1.1) is replaced by the condition that $F'(x^*; s) \ge 0$ for all s, then the natural second order sufficient conditions do not follow for general piecewise smooth functions, but only for max functions. General second order sufficient conditions can be established, but they involve conditions which are virtually impossible to test numerically. Also a complete interpretation of the multipliers in terms of the directional derivatives is available for max functions, whereas only a much weaker result holds for general piecewise smooth functions.

Thus there are intrinsic difficulties in obtaining optimality conditions, and severe limitations in designing algorithms for piecewise smooth functions more general than those expressible as a maximum of their component functions. However this latter case does include many examples of interest.

2. Piecewise smooth functions

Consider a continuous nonsmooth function $F: \mathbb{R}^n \to \mathbb{R}$ made up in some way from a finite collection of smooth (at least once continuously differentiable) functions. An example of such a function is the *composite function* defined by

$$F(x) = f(x) + h(c(x)), \tag{2.1}$$

where $f:\mathbb{R}^n \to \mathbb{R}$ and $c:\mathbb{R}^n \to \mathbb{R}^m$ are smooth and $h:\mathbb{R}^m \to \mathbb{R}$ can be nonsmooth. A particular case of interest is the *polyhedral composite function* where

$$h(c) = \max_{i=1,\dots,r} h_i^{\mathrm{T}} c + \beta_i, \tag{2.2}$$

and the vectors h_i form the columns of a matrix H. Such functions include exact penalty functions involving terms like $\|c(x)\|_p$ and $\|c(x)^+\|_p$ for $p = 1, \infty$ where $c_i^+ = \max(c_i, 0)$. Further examples of such functions and the corresponding matrices H can be found in [3]. Note that only the function h used in building up the composition is convex and that in general F is not convex.

Any polyhedral composite function can be expressed as a *finite max function*

$$F(x) = \max_{i=1,\dots,r} f_i(x), \tag{2.3}$$

by defining $f_i(x) = f(x) + h_i^{\mathrm{T}} c(x) + \beta_i$ for $i = 1, \dots, r$. Conversely any finite max function is an example of a polyhedral composite function. A slightly more general class of functions is that consisting of the functions which are *locally max functions*, that is at any point x' F can be expressed as

$$F(x) = \max_{i \in \mathcal{A}(x')} f_i(x) \tag{2.4}$$

for all x in a neighborhood of x', where

$$\mathcal{A}(x) = \{i \in 1, \dots, r \colon F(x) = f_i(x)\}. \tag{2.5}$$

Locally max functions play an important role in the following necessary and sufficient conditions.

If h is convex, then it can be expressed as a maximum (possibly infinite) of smooth functions. Although only the case where h is a polyhedral convex function is included here, corresponding to all the results that hold for locally max functions, there are results for the composite function (2.1) with h convex.

All of the above functions are locally Lipschitz and hence the generalized gradient [1] $\partial F(x)$ exists and is a nonempty convex compact set in $\mathbb{R}^n$, given by

$$\partial F(x) = \operatorname{conv}\{\nabla f_i(x) \colon i \in \mathcal{A}(x)\}. \tag{2.6}$$

These functions can also be interpreted as piecewise smooth functions made up of the smooth pieces f_i. If $\mathcal{A}(x)$ consists of a single element $\{k\}$, then F is continuously differentiable and $\partial F(x) = \nabla F(x) = \nabla f_k(x)$. It is only when $\mathcal{A}(x)$ consists of more than one index that F may be nonsmooth. Note that it is possible for $\mathcal{A}$ to consist of more than one element and for F to still be smooth. The set $\mathcal{A}$ can be thought of as the set of active pieces and there is a label (the index i) associated with each of the smooth pieces which make up the nonsmooth function. The surfaces where the pieces meet (i.e., where $\mathcal{A}(x)$ contains more than one element) are sometimes referred to as kinks.

This interpretation of a nonsmooth function as a piecewise smooth function is by no means restricted to functions made up as a polyhedral convex composition of smooth functions, or even locally max functions. To generalize this concept consider the set $\mathcal{A}$. For any $x \in \mathbb{R}^n$ the set $\mathcal{A}(x)$ is a nonempty subset of $\{1, \ldots, r\}$ satisfying

$$f_k(x) = f_i(x) \quad \forall k, i \in \mathcal{A}(x), \tag{2.7}$$

$$f_k(x) \neq f_j(x) \quad \forall k \in \mathcal{A}(x), \ \forall j \notin \mathcal{A}(x). \tag{2.8}$$

In the preceeding examples the set $\mathcal{A}$ is determined by the definition of F. Alternatively given a collection of smooth functions f_i, $i = 1, \ldots, r$ and a nonempty subset $\mathcal{A}(x)$ of $\{1, \ldots, r\}$ satisfying (2.7) and (2.8), then F can be defined from $\mathcal{A}(x)$ by

$$F(x) = f_i(x) \quad \text{for } i \in \mathcal{A}(x). \tag{2.9}$$

The smooth functions f_i are usually defined on $\mathbb{R}^n$, but this can be reduced to the requirement that each f_i is defined on an open set strictly containing the set $\{x : i \in \mathcal{A}(x)\}$. This includes a function F made up from any combination of the operations of minimizing, maximizing, or taking the modulus of the elements c_i in place of $h(c)$ in the composite function (2.1). In particular it includes functions which are not locally max functions. Note that in whichever way F is defined, the continuity of the functions f_i implies that for any point x there is an open neighborhood B of x such that

$$\mathcal{A}(y) \subset \mathcal{A}(x) \quad \text{for all } y \text{ in } B. \tag{2.10}$$

For example if $\mathcal{A}(x) = \{k\}$, then there is a neighborhood B of x such that $\mathcal{A}(y) = \{k\}$ for all y in B, as $\mathcal{A}$ is always nonempty. If $\mathcal{A}(x)$ contains more than one element then the inclusion can be strict.

An additional requirement is placed on the function F, namely that for any *directional sequence* $\{x^{(k)}\}$ converging to x in the direction s, that is $x^{(k)} = x + \alpha^{(k)} s^{(k)}$ where $\alpha^{(k)} \geq 0$, $\|s^{(k)}\| = 1$ and $s^{(k)} \to s$ as $\alpha^{(k)} \to 0$, then the following limit exists

$$\lim_{k \to \infty} \frac{F(x^{(k)}) - F(x)}{\alpha^{(k)}} = F'(x; s). \tag{2.11}$$

Here $F'(x; s)$ denotes the usual one-sided directional derivative, defined by

$$F'(x; s) = \lim_{\alpha \to 0^+} \frac{F(x + \alpha s) - F(x)}{\alpha}. \tag{2.12}$$

A function F satisfying (2.7)–(2.9) and (2.11) is referred to as a *piecewise smooth function.*

Any piecewise smooth function is locally Lipschitz so the generalized directional derivative $F^0(x; s)$ [1] is defined by

$$F^0(x; s) = \limsup_{S} \frac{F(x^{(k)} + \alpha^{(k)} s) - F(x^{(k)})}{\alpha^{(k)}}, \tag{2.13}$$

where S is the set of all sequences $\{x^{(k)}\}$ and $\{\alpha^{(k)}\}$ such that $x^{(k)} \to x$ and $\alpha^{(k)} \to 0^+$. The existence of $F'(x; s)$ implies that

$$F'(x; s) \leq F^0(x; s) \quad \forall x, s. \tag{2.14}$$

If $F'(x; s)$ exists and is equal to $F^0(x; s)$ for all x, s, then F is said to be *quasidifferentiable* [8]. For example any locally max function is quasidifferentiable, and from Demyanov and Malozemov [2]

$$F^0(x; s) = F'(x; s) = \max_{i \in \mathcal{A}(x)} s^{\mathrm{T}} \nabla f_i. \tag{2.15}$$

However for a piecewise smooth function there can exist points and directions for which $F'(x; s) < F^0(x; s)$. Also note that for any locally Lipschitz function

$$F^0(x; s) = \max_{u \in \partial F(x)} s^{\mathrm{T}} u, \tag{2.16}$$

and the condition $F^0(x; s) \geq 0$ for all s is equivalent to the condition that $0 \in \partial F(x)$. In addition, using (2.11), (2.10), (2.9), (2.6) and the smoothness of the functions f_i, one can show that a piecewise smooth function is semismooth [7]. Semismooth functions are a larger class of nonsmooth functions for which $F'(x; s)$ always exists but which are not necessarily quasidifferentiable.

A necessary condition for a point x^* to be a local minimizer of a piecewise smooth function is that

$$F'(x^*; s) \geq 0 \quad \text{for all } s, \tag{2.17}$$

which from (2.14) implies that $F^0(x^*; s) \geq 0$ for all s, or equivalently $0 \in \partial F(x^*)$. If $F'(x^*; s) = F^0(x^*; s)$ for all s (for instance if F is locally max function), then the necessary condition $0 \in \partial F^*$ is 'tight' in the sense that if $0 \in \partial F^*$, then

$$F'(x^*; s) = F^0(x^*; s) = \max_{u \in \partial F^*} s^{\mathrm{T}} u \geq 0 \quad \text{for all } s.$$

However if $F'(x^*; s) < F^0(x^*; s)$ for some s, then the necessary condition $0 \in \partial F^*$ is not 'tight' as there can exist directions s for which $F'(x^*; s) < 0$. Consider the example (Fig. 1)

$$F(x_1, x_2) = \min\{0, \max(-x_1, -x_2)\}. \tag{2.18}$$

which is the negative of an example in [7]. Then

$$0 \in \partial F(0) = \mathrm{conv}\{(0, 0)^{\mathrm{T}}, (-1, 0)^{\mathrm{T}}, (0, -1)^{\mathrm{T}}\},$$

and hence $F^0(0; s) \geq 0$ for all s. Let s be a direction with $\|s\| = 1$ and $s_2 > s_1 > 0$,

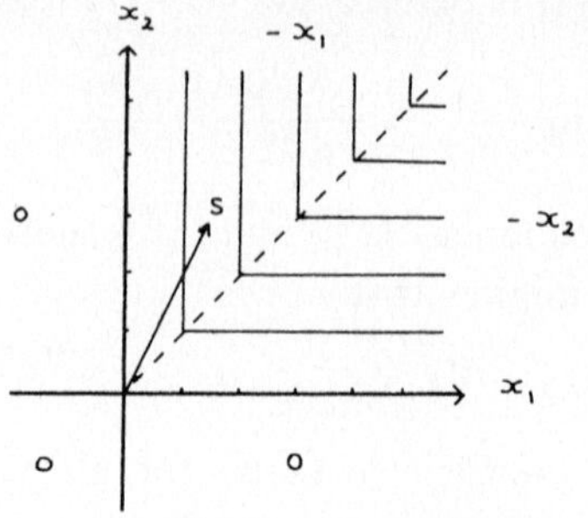

Fig. 1.

then $F'(0; s) = -s_1 < 0 \leq F^0(0; s)$. So even though $0 \in \partial F(0)$ there are directions with $F'(0; s) < 0$.

3. Optimality conditions

An interesting and informative way of looking at a piecewise smooth function is through the Lagrangian function

$$\mathscr{L}(x, \lambda) = w^T \lambda \tag{3.1}$$

where $w^T = (f_1(x), \dots, f_r(x))$. Define the set

$$\Lambda = \Lambda(x) = \left\{ \lambda \in \mathbb{R}^r : \lambda_i = 0, i \notin \mathscr{A}(x), \sum_{i \in \mathscr{A}(x)} \lambda_i = 1 \right\}. \tag{3.2}$$

Then

$$\mathscr{L}(x, \lambda) = w^T \lambda = F(x) \quad \text{for all } \lambda \in \Lambda. \tag{3.3}$$

Also from (2.10) there exists a neighborhood B of x such that

$$\Lambda(y) \subset \Lambda(x) \quad \text{for all } y \in B. \tag{3.4}$$

Let $\nabla \mathscr{L}(x, \lambda)$ denote the derivatives of the Lagrangian function with respect to the x variables only, so $\nabla \mathscr{L}(x, \lambda) = A\lambda$ where A is the matrix whose columns are the gradients $\nabla f_i(x)$, $i = 1, \dots, r$.

Theorem 3.1 (First order necessary conditions). *If x^* is a local minimizer of a piecewise smooth function, then there exist multipliers $\lambda^* \in \Lambda^*$, $\lambda^* \geq 0$ such that*

$$\nabla \mathscr{L}(x^*, \lambda^*) = A^* \lambda^* = 0. \tag{3.5}$$

Proof. As $0 \in \partial F^* = \{A^* \lambda^* : \lambda^* \in \Lambda^*, \lambda^* \geq 0\}$.

A point x' for which there exist multipliers $\lambda' \in \Lambda'$ with at least one $\lambda_i' < 0$ and

$\nabla\mathcal{L}(x', \lambda') = A'\lambda' = 0$ is referred to as a *quasi-stationary point*. The multipliers $\lambda \in \Lambda$ can be regarded as the derivatives of the nonsmooth function F with respect to the component functions f_i. Hence at a quasi-stationary point the $\lambda_i' < 0$ implies that F can be reduced by moving in a direction s such that $i \notin \mathcal{A}(x' + \alpha s)$ (see Theorem 4.1 later in the paper), in other words removing the piece i from the definition of F.

If F is convex, then F is a locally max function and the conditions of Theorem 3.1 are necessary and sufficient. In general however a piecewise smooth function is not convex and it is the curvature in the directions s for which $F'(x^*; s) = 0$ which characterize a local minimum. Before second order conditions can be discussed the relationship between two sets must be considered. The following treatment of second order conditions closely follows that given by Fletcher and Watson [4] for the case when $F(x) = f(x) + \|c(x)\|$. Let

$$G^* = \{s \colon F'(x^*; s) = 0, \|s\| = 1\} \tag{3.6}$$

be the set of directions of zero slope at the point x^*. If x^* is a local minimizer of F, then G^* contains the set

$$G_0^* = \{s \colon F^0(x^*; s) = 0\}, \tag{3.7}$$

as $s \in G_0^*$ implies $0 \le F'(x^*; s) \le F^0(x^*; s) = 0$, so $s \in G^*$. However if x' is a point for which $0 \in \partial F'$, then G_0' may not be a subset of G'. Define the set X by

$$X = \{x \colon F(x) = w^{\mathrm{T}}\lambda^*\} \tag{3.8}$$

where λ^* is one of the vectors that exist in Theorem 3.1. The set X contains the set of points $\{x \colon \mathcal{A}(x) = \mathcal{A}^*\}$ as then $\lambda^* \in \Lambda^* = \Lambda$ and $F(x) = \mathcal{L}(x, \lambda^*) = w^{\mathrm{T}}\lambda^*$ by (3.3). This inclusion can be strict if $\lambda_i^* = 0$ for some $i \in \mathcal{A}^*$. Now let $\mathcal{G}^*$ be the set of feasible directions for X, that is $s \in \mathcal{G}^*$ implies there exists a directional sequence $\{x^{(k)}\}$ feasible in X. It is the relationship between the sets $\mathcal{G}^*$ and G^* that is of primary interest.

Lemma 3.1. $\mathcal{G}^* \subset G^*$.

Proof. Let $s \in \mathcal{G}^*$, then there exists a directional sequence $\{x^{(k)}\}$ converging to x^* in the direction s, which is feasible in X such that

$$F'(x^*; s) = \lim_{k\to\infty} \frac{F^{(k)} - F^*}{\alpha^{(k)}} = \lim_{k\to\infty} \frac{w^{(k)\mathrm{T}}\lambda^* - w^{*\mathrm{T}}\lambda^*}{\alpha^{(k)}} = s^{\mathrm{T}}A^*\lambda^* = 0,$$

by (3.3), (3.8), (3.5) and the Taylor series expansion

$$w^{(k)} = w^* + A^{*\mathrm{T}}s + o(\alpha^{(k)}).$$

Thus $s \in G^*$.

The inclusion the other way may not hold and such cases are discussed later.

For now the regularity assumption

$$G^* = \mathcal{G}^* \tag{3.9}$$

is used to get second order necessary conditions. The validity of this assumption depends on the multipliers λ^* which exist in Theorem 3.1. In the following each of the component functions f_i must be at least twice continuously differentiable.

Theorem 3.2 (Second order necessary conditions). *If x^* is a local minimizer of a piecewise smooth function F, then by Theorem 3.1 there exists a $\lambda^* \in \Lambda^*$, $\lambda^* \geq 0$ such that $A^*\lambda^* = 0$. If in addition $\mathcal{G}^* = G^*$, then*

$$s^{\mathrm{T}}\nabla^2\mathcal{L}(x^*, \lambda^*)s \geq 0 \quad \forall s \in G^*. \tag{3.10}$$

Proof. As $\mathcal{G}^* = G^*$ any $s \in G^*$ is in $\mathcal{G}^*$, so there exists a directional sequence $\{x^{(k)}\}$ converging to x^* in the direction s which is feasible in X. By a Taylor series expansion

$$\mathcal{L}(x^{(k)}, \lambda^*) = \mathcal{L}(x^*, \lambda^*) + \alpha^{(k)}s^{(k)\mathrm{T}}\nabla\mathcal{L}(x^*, \lambda^*)$$
$$+ \tfrac{1}{2}\alpha^{(k)2}s^{(k)\mathrm{T}}\nabla^2\mathcal{L}(x^*, \lambda^*)s^{(k)} + o(\alpha^{(k)2})$$

where $x^{(k)} - x^* = \alpha^{(k)}s^{(k)}$. From (3.5) and (3.8) this gives

$$F^{(k)} = F^* + \tfrac{1}{2}\alpha^{(k)2}s^{(k)\mathrm{T}}\nabla^2\mathcal{L}(x^*, \lambda^*)s^{(k)} + o(\alpha^{(k)2}).$$

As F^* is a local minimum, $F^{(k)} \geq F^*$ for all k sufficiently large, so dividing by $\tfrac{1}{2}\alpha^{(k)2}$ and taking the limit gives (3.10) as desired.

Second order sufficient conditions depend on whether F is a locally max function or not. One sufficient condition is that

$$F'(x^*; s) > 0 \quad \text{for all } s, \tag{3.11}$$

which from (2.14) gives $F^0(x^*; s) > 0$ or equivalently $0 \in \operatorname{int} \partial F^*$. The condition $0 \in \operatorname{int} \partial F^*$ is only sufficient when F is a locally max function or more generally when F is quasidifferentiable. Also as $\partial F^* = \operatorname{conv}\{\nabla f_i^*: i \in \mathcal{A}^*\}$ $0 \in \operatorname{int} \partial F^*$ implies that $\mathcal{A}^*$ has at least $n+1$ elements. Even when $\mathcal{A}^*$ has $n+1$ or more elements there can exist directions for which $F'(x^*; s) = 0$. This also occurs whenever $\mathcal{A}^*$ has fewer than $n+1$ elements. In these cases it is again the curvature in these directions which is important. $F'(x^*; s) > 0$ for all s implies that G^* is empty, and hence $\mathcal{G}^* = G^*$ by Lemma 3.1. The following theorem gives second order sufficient conditions when G^* is nonempty for piecewise smooth functions which are also locally max functions. No regularity assumption is needed.

Theorem 3.3 (Second order sufficient conditions, locally max functions). *Let F be a piecewise smooth function. If x^* is a point about which F is locally a finite* *max function and if there exists a vector* λ^* *satisfying*

$$\lambda^* \geq 0, \qquad \lambda^* \in \Lambda^* \quad \textit{and} \quad A^*\lambda^* = 0 \tag{3.12}$$

such that

$$s^{\mathrm{T}}\nabla^2 \mathscr{L}(x^*, \lambda^*)s > 0 \quad \forall s \in G^*, \tag{3.13}$$

then x^* *is a unique local minimizer of* F.

Proof. For a proof by contradiction assume there exists a sequence $\{x^{(k)}\}$ converging to x^* such that $F^{(k)} \leq F^*$ for all k. Then there is a subsequence which is a directional sequence such that $x^{(k)} = x^* + \alpha^{(k)}s^{(k)}$, $\alpha^{(k)} \geq 0$, $\|s^{(k)}\| = 1$ and $s^{(k)} \to s$ as $\alpha^{(k)} \to 0$. Then as $F^{(k)} \leq F^*$ for all k

$$0 \geq \lim_{k\to\infty} \frac{F^{(k)} - F^*}{\alpha^{(k)}} = F'(x^*; s). \tag{3.14}$$

Also as F is locally expressible as a finite max function

$$F'(x^*; s) = F^0(x^*; s) = \max_{u \in \partial F^*} s^{\mathrm{T}}u = \max_{\lambda \in \Lambda^*, \lambda \geq 0} s^{\mathrm{T}}A^*\lambda \geq 0$$

by (3.12). Combining these two inequalities gives $F'(x^*; s) = 0$ or $s \in G^*$. Now as F is a finite max function in some neighborhood about x^*, for k sufficiently large $w^{(k)\mathrm{T}}\lambda^{(k)} \geq w^{(k)\mathrm{T}}\lambda^*$ for any $\lambda^{(k)} \in \Lambda^{(k)}$, $\lambda^{(k)} \geq 0$ by (3.4). Then

$$\begin{aligned} 0 \geq F^{(k)} - F^* &= w^{(k)\mathrm{T}}\lambda^{(k)} - w^*\lambda^* \qquad \text{(for any } \lambda^{(k)} \in \Lambda^{(k)}, \lambda^{(k)} \geq 0) \\ &\geq w^{(k)\mathrm{T}}\lambda^* - w^*\lambda^* \\ &= \mathscr{L}(x^{(k)}, \lambda^*) - \mathscr{L}(x^*, \lambda^*) \\ &= \tfrac{1}{2}\alpha^{(k)2}s^{(k)\mathrm{T}}\nabla^2\mathscr{L}(x^*, \lambda^*)s^{(k)} + o(\alpha^{(k)2}), \end{aligned}$$

by a Taylor series expansion and (3.12). Dividing through by $\frac{1}{2}\alpha^{(k)2}$ and taking the limit gives a contradiction to (3.13).

The restriction to piecewise smooth functions locally expressible as a finite max function about x^* is essential as the following example illustrates (Fig. 2). Let $F: \mathbb{R} \to \mathbb{R}$ be defined by

$$F(x) = \begin{cases} f_1(x) = x^2, & x \geq 0, \\ f_2(x) = x^3, & x \leq 0, \end{cases}$$

and let $x^* = 0$. Then $\partial F^* = \mathrm{conv}\{\nabla f_1^*, \nabla f_2^*\} = \{0\}$ and λ^* in Theorem 3.3 can be any vector $\lambda^* \in \mathbb{R}^2$ such that $\lambda^* \geq 0$ and $\lambda_1^* + \lambda_2^* = 1$. Then $G^* = \{1, -1\}$ and for any $s \in G^*$, $s^{\mathrm{T}}\nabla^2\mathscr{L}(x^*, \lambda^*)s = 2\lambda_1^* > 0$ for all possible choices of λ^* except $\lambda_1^* = 0$, $\lambda_2^* = 1$. Yet clearly x^* is not a local minimizer of F. Note that the function F is not expressible as the maximum of f_1 and f_2 in any neighborhood of 0. The trouble in this case does not arise from the weakness in the necessary

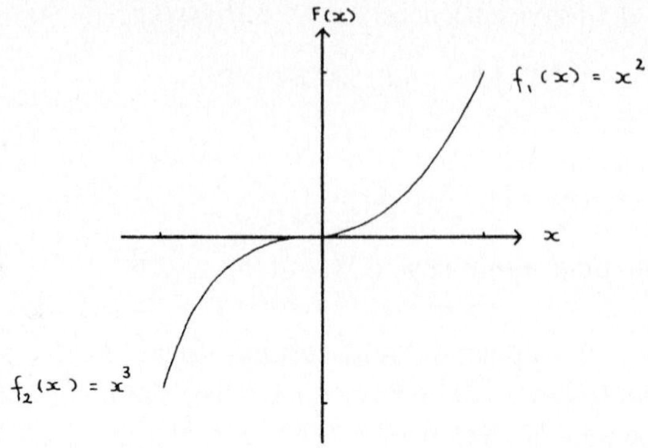

Fig. 2.

conditions $0 \in \partial F^*$ as $F'(x^*; s) \geq 0$ for all s, but from the multiple choices of λ^*. To get a sufficiency theorem for general piecewise smooth functions all the possible λ^* must be considered and the weakness in the necessary conditions $0 \in \partial F^*$ overcome.

Theorem 3.4 (Second order sufficient conditions). *Let F be a piecewise smooth function and let x^* be a point such that $F'(x^*; s) \geq 0$ for all s. Let $\Gamma^* = \{\lambda : \lambda \in \Lambda^*, \lambda \geq 0\}$. If for all $\lambda^* \in \Gamma^*$, and all $s \in G^*$*

$$s^{\mathrm{T}} \nabla^2 \mathscr{L}(x^*, \lambda^*) s > 0, \tag{3.15}$$

then x^* is a unique local minimizer of F.

Proof. The proof is along the same lines as that to Theorem 3.3. Assume there exists a sequence and hence a directional sequence $\{x^{(k)}\} \to x^*$ in the direction s such that $F^{(k)} \leq F^*$ for all k. This implies that

$$0 \geq \lim_{k \to \infty} \frac{F^{(k)} - F^*}{\alpha^{(k)}} = F'(x^*; s),$$

which together from the conditions on x^* implies $F'(x^*; s) = 0$ or $s \in G^*$. Now from (3.4) $\Lambda^{(k)} \subset \Lambda^*$ so, by considering subsequences if necessary, there exists a $\lambda' \geq 0$ such that $\lambda' \in \Lambda^{(k)} \subset \Lambda^*$ for all such k. By (3.3)

$$\begin{aligned} 0 \geq F^{(k)} - F^* &= w^{(k)\mathrm{T}} \lambda' - w^{*\mathrm{T}} \lambda' \\ &= \alpha^{(k)} s^{(k)\mathrm{T}} A^* \lambda' + \tfrac{1}{2} \alpha^{(k)2} s^{(k)\mathrm{T}} \nabla^2 \mathscr{L}(x^*, \lambda') s^{(k)} + o(\alpha^{(k)2}). \end{aligned}$$

Now $F'(x^*; s) = 0$ implies that $s^{\mathrm{T}} A^* \lambda' = 0$, so dividing by $\frac{1}{2}\alpha^{(k)2}$ and taking the limit implies

$$0 \geq s^{\mathrm{T}} \nabla^2 \mathscr{L}(x^*, \lambda') s \quad \text{for some } \lambda' \in \Gamma^*,$$

which contradicts (3.15).

The problem is that the theorem contains two conditions which in their current form are not numerically verifiable. Firstly the condition $F'(x^*; s) \geq 0$ for all s must be used to overcome the weakness in the necessary condition $0 \in \partial F^*$ for piecewise smooth functions which are not locally max functions. Secondly (3.15) must hold for all $\lambda^* \in \Gamma^*$ rather than just one $\lambda^* \in \Gamma^*$ such that $A^*\lambda^* = 0$. In the latter case a λ^* can be calculated and so some attempt to verify (3.15) can be made. It may be necessary only to consider all the $\lambda^* \in \Gamma^*$ such that $A^*\lambda^* = 0$, however the above proof does not hold as λ' may not satisfy $A^*\lambda' = 0$.

4. Discussion

The usual necessary conditions $0 \in \partial F^*$ can be 'tightened' for piecewise smooth functions which are not locally max functions by replacing the condition of the existence of a vector λ^* satisfying

$$\lambda^* \in \Lambda^*, \qquad \lambda^* \geq 0, \qquad A^*\lambda^* = 0 \tag{4.1}$$

by the conditions

$$F'(x^*; s) \geq 0 \quad \text{for all } s. \tag{4.2}$$

Given the set $\mathcal{A}^*$ and the matrix A^* the existence of a vector λ^* satisfying (4.1) can be checked computationally. The problem of finding a λ^* satisfying (4.1), or, if that is impossible, finding the closest approximation to such a λ^* in some sense, is often used as a basis for algorithms for nonsmooth optimization (for example finding the closest point to the origin in the set $\partial F^* = \{A^*\lambda : \lambda \in \Lambda^*, \lambda \geq 0\}$ in the Euclidean norm [9]). The hope is then that the conditions (4.1) are almost sufficient so that a point for which (4.1) is satisfied is a local minimum. However if F is not a locally max function, then there are functions for which (4.1) holds but (4.2) does not. If F is convex, then it can be expressed as a maximum (possibly infinite) of smooth functions and in this case (4.1) and (4.2) are equivalent. The conditions (4.1) and (4.2) are also equivalent if F is quasidifferentiable. The conditions (4.2) have the severe disadvantage of being virtually impossible to check computationally.

If the regularity assumption $\mathcal{G}^* = G^*$ holds, then the second order necessary conditions apply to general piecewise smooth functions.

Let λ^* satisfy (4.1) and define

$$\mathcal{A}^*_+ = \{i \in \mathcal{A}^* : \lambda^*_i > 0\}, \tag{4.3}$$

and let $q \in \mathcal{A}^*_+$. Also let N^* be the matrix whose columns are the vectors $\nabla f^*_q - \nabla f^*_i$ for $i \in \mathcal{A}^* - q$. If $F'(x^*; s) > 0$ for all s, then $G^* = \emptyset$, and by Lemma 3.1 $\mathcal{G}^* = G^*$. For locally max functions this holds if there is a subset I^* of $\mathcal{A}^*$ containing $n+1$ elements such that the corresponding matrix N^* has full rank and such that a λ^* satisfying (4.1) has $\lambda^*_i > 0$ for all $i \in I^*$, as in this case

$0 \in \operatorname{int} \partial F^*$. More generally if the matrix N^* has full rank, then $\mathcal{G}^* = G^*$, for locally max functions. This result is essentially contained in [4]. The proof depends on the fact that

$$F'(x^*; s) = F^0(x^*; s) = \max_{\lambda \in \Lambda^*, \lambda \geq 0} s^{\mathrm{T}} A^* \lambda \tag{4.4}$$

and on the locally max nature of F. Whether a corresponding result holds for general piecewise smooth functions is uncertain. The linear independence condition cannot be satisfied if $\mathcal{A}^*$ contains more than $n+1$ elements (for example $F(x) = \|c(x)\|_1$ where there are 2^t active pieces at x^* where t is the number of zero elements of $c(x^*)$). This case however can be treated in an alternate way (see [4]).

When $\mathcal{A}^*$ contains $t < n+1$ elements such that the matrix N^* with columns $\nabla f_q^* - \nabla f_i^*$, $i \in \mathcal{A}^* - q$ where $q \in \mathcal{A}_\ddagger^*$ has full rank, then the equations

$$f_q(x) - f_i(x) = 0, \quad i \in \mathcal{A}^* - q \tag{4.5}$$

determine an $n+1-t$ dimensional surface on which x^* must lie. It is then the curvature within this surface that determines the minimum. If $\mathcal{A}^*$ contains $n+1$ elements (or there is a subset of $\mathcal{A}^*$ with $n+1$ elements) such that the corresponding matrix N^* has full rank, then x^* is uniquely determined by the equations (4.5).

The following theorem clarifies the interpretation of the multipliers λ_i as the derivatives of the nonsmooth function F with respect to the component functions f_i for $i \in \mathcal{A}$.

Theorem 4.1. *Let F be a locally max function, and let x' be a point where $\mathcal{A}'$ has more than one element such that there exists a vector λ' with*

$$\lambda' \in \Lambda' \quad \text{and} \quad A'\lambda' = 0. \tag{4.6}$$

Let t denote the number of elements in $\mathcal{A}'$, and let N' be the matrix with columns $\nabla f_q' - \nabla f_i'$ for $i \in \mathcal{A}' - q$ where $q \in \mathcal{A}'$.

If the vectors

$$\begin{bmatrix} 1 \\ \nabla f_i' \end{bmatrix}$$

for $i \in \mathcal{A}'$ are linearly independent, then λ' is the unique solution to

$$N'\mu = \nabla f_q' \tag{4.7}$$

recovered by

$$\lambda_i' = \mu_i \quad \text{for } i \in \mathcal{A}' - q, \qquad \lambda_q' = 1 - \sum_{i \in \mathcal{A}' - q} \lambda_i',$$

and

(1) *If $\lambda_i' \geq 0$ for all $i \in \mathcal{A}'$, then $F'(x'; s) \geq 0$ for all s, and*

(a) *if* $t = n+1$ *and* $\lambda_i' > 0$ *for all* $i \in \mathcal{A}'$, *then* $F'(x'; s) > 0$ *for all* s *with* $\|s\| = 1$.

(b) *if* $t < n+1$ *and* $\lambda_i' > 0$ *for all* $i \in \mathcal{A}'$, *then* $F'(x'; s) = 0$ (*where* $\|s\| = 1$) *if and only if* s *is tangential to the surface*

$$\{x : f_i(x) = f_j(x),\ \forall i, j \in \mathcal{A}'\} \tag{4.8}$$

at x'.

(c) *if there exists a* $k \in \mathcal{A}'$ *such that* $\lambda_k' = 0$, *then any direction* s *tangential to the surface*

$$\{x : f_i(x) = f_j(x),\ \forall i, j \in \mathcal{A}_+'\} \tag{4.9}$$

at x' *and with* $s^{\mathrm{T}} \nabla f_i' \le 0$ *for all* $i \in \mathcal{A}' - \mathcal{A}_+'$ *has* $F'(x'; s) = 0$.

(2) *If there exists a* $k \in \mathcal{A}'$ *such that* $\lambda_k' < 0$, *then any direction* s *with* $\|s\| = 1$ *and tangential to the surface*

$$\{x : f_i(x) = f_j(x),\ \forall i, j \in \mathcal{A}' - k\} \tag{4.10}$$

at x' *and with* $s^{\mathrm{T}} \nabla f_i' < 0$ *for some* $i \in \mathcal{A}' - k$ *has* $F'(x'; s) < 0$, *moreover such a direction* s *always exists.*

Proof. It can readily be shown that the vectors

$$\begin{bmatrix} 1 \\ \nabla f_i' \end{bmatrix}$$

for $i \in \mathcal{A}'$ are linearly independent if and only if the vectors $\nabla f_q' - \nabla f_i'$ for $i \in \mathcal{A}' - q$ are linearly independent (for any $q \in \mathcal{A}'$), so N' has full rank. The equivalence of (4.6) and (4.7) simply comes from using the variable λ_q' to eliminate the constraint in (3.2). The vector λ' is unique as N' has full rank.

For locally max functions (2.15) holds. Thus the first part of (1) comes from the equivalence of $F^0(x; s) \ge 0$ for all s and

$$0 \in \partial F' = \{A\lambda : \lambda \in A', \lambda \ge 0\}. \tag{4.11}$$

The conditions of part (a) imply that $0 \in \operatorname{int} \partial F'$ and hence that $F'(x'; s) > 0$ for all s with $\|s\| = 1$. For the proof of part (b) let $s \in G'$ (defined in (3.6)), then as F is a locally max function

$$0 = F'(x'; s) = \max_{\lambda \in \Lambda', \lambda \ge 0} s^{\mathrm{T}} A' \lambda = \max_{i \in \mathcal{A}'} s^{\mathrm{T}} \nabla f_i' \tag{4.12}$$

and the maximum is achieved by λ'. As $\lambda_i' > 0$ for all $i \in \mathcal{A}'$ this implies $s^{\mathrm{T}} \nabla f_i' = 0$ for all $i \in \mathcal{A}'$ and hence $s^{\mathrm{T}} N' = 0$. This is only possible for $t < n+1$, otherwise N' being full rank would contradict $\|s\| = 1$. The surface (4.8) can also be described by

$$\{x : f_q(x) - f_i(x) = 0,\ \forall i \in \mathcal{A}' - q\},$$

so the columns of N' are the normals to the equations defining the surface at x', which gives the desired result. Conversely let s be a direction with $\|s\| = 1$ and tangential to the surface (4.8) at x'. Then $s^T\nabla f'_i = s^T\nabla f'_j$ for all i, j in $\mathscr{A}'$ and hence for any $j \in \mathscr{A}'$

$$s^T\nabla f'_j = \sum_{i\in\mathscr{A}'} \lambda'_i s^T\nabla f'_i = s^T A'\lambda' = 0 \tag{4.13}$$

and hence $F'(x'; s) = 0$ from (2.15).

For part (c) let s be tangential to the surface (4.9) at x' so

$$s^T\nabla f'_i = s^T\nabla f'_j, \quad \forall i, j \in \mathscr{A}'_+.$$

As $\lambda'_i = 0$ for all $i \in \mathscr{A}' - \mathscr{A}'_+$ the result follows from (4.13) and (2.15).

For part (2) let s be a vector with $\|s\| = 1$ satisfying the conditions described, so that

$$s^T\nabla f'_i = s^T\nabla f'_j, \ \forall i, j \in \mathscr{A}' - k.$$

Then it follows from (4.6) that

$$s^T\nabla f'_k = \frac{(\lambda'_k - 1)}{\lambda'_k} s^T\nabla f'_i \quad \text{for any } i \in \mathscr{A}' - k.$$

Now $\lambda'_k < 0$ implies $(\lambda'_k - 1)/\lambda'_k > 1$ so $F'(x'; s) < 0$ from (2.15).

It now only remains to show that vectors satisfying the conditions of part (2) can always be found. One way this can be done is to note that if there is no s satisfying

$$s^T\nabla f'_i < 0 \quad \text{for all } i \in \mathscr{A}',$$

then by Gordan's Theorem of the alternative [6] the system

$$A'\lambda = 0, \tag{4.14}$$

$$\lambda \geq 0 \tag{4.15}$$

has a nonzero solution. This solution can always be scaled to satisfy

$$\sum_{i\in\mathscr{A}'} \lambda_i = 1. \tag{4.16}$$

However from the linear independence condition λ' is the unique solution to (4.14) plus (4.16), so $\lambda'_k < 0$ gives a contradiction to (4.15).

If F is not a locally max function, then part (1) of Theorem 4.1 no longer holds. However for any piecewise smooth function

$$F'(x; s) \leq \max_{i\in\mathscr{A}(x)} s^T\nabla f_i(x),$$

so part (2) of the theorem still holds.

Let π denote the surface $\{x: f_q(x) - f_i(x) = 0, \forall i \in \mathscr{A}^* - q\}$ where $q \in \mathscr{A}^*$ and assume N^* has full rank. The columns of N^* are normals to the surface π at x^*. Let $\mathcal{A}^*$ have t elements and let Z^* be an $n \times n+1-t$ orthogonal matrix such that $Z^{*\mathrm{T}}N^* = 0$ (such a matrix can always be found as N^* has full rank; for example by a QR decomposition). Then the columns of Z^* span the space orthogonal to the columns of N^*, that is the space tangential to the surface π at x^*. If $t = n+1$, then the surface π consists of the single point x^* and Z^* does not exist. If $t = 1$, then there is no surface π, which is the case in normal smooth optimization problems. From Theorems 3.2 and 4.1 a necessary condition for a local minimum is that in addition to (3.5) the matrix

$$W_R^* = Z^{*\mathrm{T}}\nabla^2\mathcal{L}(x^*, \lambda^*)Z^*$$

is positive semi-definite. This also holds for general piecewise smooth functions as a direction s tangential to π at x^* is in $\mathcal{G}^*$. If in addition the multipliers λ_i^* are strictly positive for all $i \in \mathcal{A}^*$, then a sufficient condition for locally max functions is that (3.5) holds and W_R^* is positive definite. The surface π is a surface of nondifferentiability of the function F and W_R^* can be interpreted as the curvature of the function F within the surface π at x^*.

Acknowledgment

I am indebted to R. Fletcher for his advice and many helpful discussions and to one of the referees for his useful comments.

References

[1] F.H. Clarke, "Generalized gradients and applications", *Transactions of the American Mathematical Society* 205 (1975) 247–262.
[2] V.F. Dem'yanov and V.N. Malozemov, *Introduction to minimax* (Wiley, New York, 1974).
[3] R. Fletcher, "A model algorithm for composite NDO problems", presented at the Workshop on Numerical Techniques in System Engineering, University of Kentucky, Lexington, KY, June 16–20 (1980).
[4] R. Fletcher and G.A. Watson, "First and second order conditions for a class of nondifferentiable optimization problems", *Mathematical Programming* 18 (1980) 291–307.
[5] J.-B. Hiriart-Urruty, "On optimality conditions in mathematical programming", *Mathematical Programming* 14 (1978) 73–86.
[6] O.L. Mangasarian, *Nonlinear programming* (McGraw-Hill, New York, 1969).
[7] R. Mifflin, "Semismooth and semiconvex functions in constrained optimization", *SIAM Journal on Control and Optimization* 15 (1977) 959–972.
[8] V.N. Pshenichyni, *Necessary conditions for an extremum* (Dekker, New York, 1971).
[9] P. Wolfe, "A method of conjugate subgradients for minimizing nondifferentiable functions", *Mathematical Programming Study* 3 (1975) 145–173.

Mathematical Programming Study 17 (1982) 28–66.
North-Holland Publishing Company

LAGRANGE MULTIPLIERS AND SUBDERIVATIVES OF OPTIMAL VALUE FUNCTIONS IN NONLINEAR PROGRAMMING*

R.T. ROCKAFELLAR

Department of Mathematics, University of Washington, Seattle, WA 98195, U.S.A.

Received 24 September 1980
Revised manuscript received 25 March 1981

For finite-dimensional optimization problems with locally Lipschitzian equality and inequality constraints and also an abstract constraint described by a closed set, a Lagrange multiplier rule is derived that is sharper is in some respects than the ones of Clarke and Hiriart-Urruty. The multiplier vectors provided by this rule are given meaning in terms of the generalized subgradient set of the optimal value function in the problem with respect to perturbational parameters. Bounds on subderivatives of the optimal value function are thereby obtained and in certain cases the existence of ordinary directional derivatives.

Key words: Lagrange Multipliers, Subgradients, Marginal Values, Nonlinear Programming.

1. Introduction

In this paper we study an optimization problem that depends on parameter vectors $u = (u_1, \dots, u_m) \in \mathbb{R}^m$ and $v = (v_1, \dots, v_d) \in \mathbb{R}^d$:

$$(\mathrm{P}_{u,v}) \qquad \begin{array}{l} \text{minimize} \quad f_0(v, x) \text{ over all } x \in \mathbb{R}^n \text{ such that } (v, x) \in D \text{ and} \\ f_i(v, x) + u_i \begin{cases} \leq 0 & \text{for } i = 1, \dots, s, \\ = 0 & \text{for } i = s+1, \dots, m, \end{cases} \end{array}$$

where D is a subset of $\mathbb{R}^d \times \mathbb{R}^n$ and each f_i is a real-valued function on an open set which includes D. We assume that D is closed and f_i is locally Lipschitzian on D (i.e., Lipschitz continuous relative to some $\mathbb{R}^n$-neighborhood of each point of D). Examples where such assumptions are fulfilled include

(a) *the smooth case*: $D = \mathbb{R}^d \times \mathbb{R}^n$ and every f_i of class $\mathscr{C}^1$;

(b) *the convex case*: D closed convex, f_i convex for $i = 0, 1, \dots, s$ and affine for $i = s+1, \dots, m$;

(c) *the mixed smooth-convex case*: $D = \mathbb{R}^d \times C$ with C closed convex, $f_i(v, x)$ of class $\mathscr{C}^1$ with respect to v (the gradient depending continuously on (v, x) rather than just v), as well as convex in x for $i = 0, 1, \dots, s$ and affine in x for $i = s+1, \dots, m$.

Clarke [3] has obtained a Lagrange multiplier rule that unifies the known

*Research sponsored by the Air Force Office of Scientific Research, Air Force Systems Command, USAF, under grant no. 77–3204 at the University of Washington, Seattle.

first-order necessary conditions for optimality in nonlinear programming problems of types (a), (b) or (c) but is applicable as well to the general case of $(P_{u,v})$. This rule has been sharpened by Hiriart-Urruty [18]. Our objective here is to explore the connection between the Lagrange multipliers for the constraints in $(P_{u,v})$ as provided by such a rule, and certain generalized directional derivatives and subgradients of the function

$$p(u, v) = \inf(P_{u,v}) \qquad \text{(global optimal value)}$$

with respect to both u and v.

The function p is everywhere-defined on $\mathbb{R}^m \times \mathbb{R}^d$ under the convention that $p(u, v) = +\infty$ when $(P_{u,v})$ is infeasible, but it can well fail to be differentiable in the ordinary sense at points where it is finite, even in the smooth case (a). Nevertheless, p is of such obvious interest that quite apart from any connection with Lagrange multipliers, there is strong motivation for pushing beyond differentiability to some sort of subdifferential theory of its properties. Generalized derivatives of p have direct significance in sensitivity analysis and in determining criteria for Lipschitzian behavior of p and the like. They also furnish information that might be used in minimizing $p(u, v)$ subject to further constraints on u and v, as can be the task posed in decomposition techniques where $(P_{u,v})$ appears as just a subproblem of a larger problem.

Of even greater importance, though, is the role that generalized derivatives of the optimal value function p can have in answering fundamental questions about the existence, uniqueness and interpretation of Lagrange multiplier vectors, questions which have a bearing on many aspects of theory and computation. This role is well understood in the convex case (cf. [26]) and to some extent also through partial results in the smooth and mixed cases listed above, but it has not been clarified for $(P_{u,v})$ in general.

Roughly speaking on the basis of experience in the special cases which have been tackled, possible rates of change of $p(u, v)$ with respect to u_i should have something to do with possible multiplier values y_i associated with the ith constraint in $(P_{u,v})$. The study of variations with respect to the parameters v_l as well as u_i is approachable by the same idea, because $p(u, v)$ can equally be regarded as the optimal value in the problem:

$$(P'_{u,v}) \qquad \begin{aligned} &\text{minimize} \quad f_0(w, x) \text{ over all } (w, x) \in D \text{ satisfying} \\ &\qquad f_i(w, x) + u_i \begin{cases} \leq 0 & \text{for } i = 1, \dots, s, \\ = 0 & \text{for } i = s+1, \dots, m \end{cases} \\ &\qquad -w_l + v_l = 0 \quad \text{for } l = 1, \dots, d. \end{aligned}$$

Multipliers z_l associated with the constraint $-w_l + v_l = 0$ in $(P'_{u,v})$ should be related to some kind of derivative of $p(u, v)$ with respect to v_l, but in view of the equivalence between $(P'_{u,v})$ and $(P_{u,v})$, such multipliers are bound to have close ties with the multipliers y_i.

Altogether then, a duality may be expected between Lagrange multiplier vectors for the constraints in $(P_{u,v})$ and subdifferential properties of $p(u, v)$. Insofar as this can be formalized, it should afford valuable insight in both directions. The development of a really far-reaching duality beyond the convex case has been hampered, however, by a lack of appropriate mathematical tools and concepts.

Most of the past work on subdifferential properties of the function p has gone into the determination of formulas for the one-sided directional derivatives

$$p'(u, v; h, k) = \lim_{t \downarrow 0} \frac{p(u+th, v+tk) - p(u, v)}{t} \tag{1.1}$$

or bounds on the corresponding upper or lower Dini derivatives, where 'lim' is replaced by 'lim sup' and 'lim inf'. In the convex case (b), $p(u, v)$ is actually convex in (u, v), and $p'(u, v; h, k)$ exists for every (h, k) [26, Sections 28–29]. A theorem of Gol'shtein [15] shows that $p'(u, v; h, k)$ also exists in the mixed case (c) when the set of saddlepoints of the Lagrangian in $(P_{u,v})$ is nonempty and bounded. This result, proved independently by Hogan [20], generalizes the Mills–Williams marginal value theorem in linear programming [33]. Dini derivatives were studied by Gauvin and Tolle [13] in the smooth case (a) and by Auslender [2] in the somewhat more general situation where only the equality constraints in (a) are $\mathscr{C}^1$. Bounds on Dini derivatives were used by Gauvin and Tolle to demonstrate the existence of $p'(u, v; h, k)$ under certain circumstances [13] and by Gauvin [11] to get a criterion for p to be locally Lipschitzian in the smooth case. The cited results of Gauvin and Tolle [13], Auslender [2] and Gauvin [11], ostensibly treat only parameters of type u_i, but they can be extended to parameters of type v_l using the reformulation of $(P_{u,v})$ as $(P'_{u,v})$. For a direct approach to such parameters, cf. [12] and related work of Fiacco and Hutzler [10].

The infinite-dimensional case too has been studied to a certain extent [21–23, 14]. Gollan [14] gives his own definition of Lagrange multipliers for nonsmooth problems, quite different from the Lagrange multipliers of Clarke mentioned earlier, but when his results are applied to classical cases they do not yield derivative bounds as strong as those of Gauvin and Tolle, for instance. Other work on ordinary one-sided derivatives of optimal value functions that should be noted for exceeding the framework in this paper in some respects, although involving significant restrictions in others, is that of Dem'janov et al. [7, 8].

Our objective here is to explore the subdifferential properties of the function p, including extensions of the results cited above, by means of a broader kind of nonlinear analysis that has blossomed from ideas of Clarke [4]. This method of analysis, the pertinent parts of which will be reviewed in Section 2, deals with certain generalized subgradients of p and corresponding 'subderivatives' that are more suited in some ways to the description of functions as irregular as p can be. Smoothness or convexity assumptions on $(P_{u,v})$ are not required, yet the theory is such that the consequences of such assumptions are readily ascertained.

Subdifferential analysis in this sense has already been applied to optimal value functions like p, although not in such a thorough-going manner as in the present contribution. Clarke himself has employed a mild subderivative condition on p called 'calmness' as a constraint qualification in the derivation of his Lagrange multiplier rule [3]. A result of Gauvin [11] furnishes an outer estimate for the subgradient set $\partial p(u, v)$ in the smooth case (a). This has been carried to certain nonsmooth cases of $(P_{u,v})$, but with smooth equality constraints, by Hiriart-Urruty [7] as part of a more abstract study of marginal values. Clarke and Aubin [6] and Aubin [1] have established for other special cases of $(P_{u,v})$, via some theorems in a Banach space setting accompanied by a number of convexity assumptions, the existence in $\partial p(u, v)$ of certain multiplier vectors—thus, 'inner estimates' for $\partial p(u, v)$. All these results have concerned situations where p is Lipschitzian in a neighborhood of (u, v), and the authors (except for Hiriart-Urruty) have provided conditions on $(P_{u,v})$ that ensure this Lipschitzian behavior. In contrast, exact formulas for $\partial p(u, v)$ in the general case of $(P_{u,v})$ that are valid whether or not p is locally Lipschitzian have been given by Rockafellar [29], but in terms of *limits* of sequences of special multiplier vectors corresponding to saddle-points of the augmented Lagrangian in neighboring problems (P_{u^j, v^j}).

In this paper we derive inner and outer estimates for $\partial p(u, v)$ in terms of Lagrange multiplier vectors that satisfy Clarke's necessary conditions for $(P_{u,v})$ itself (see Section 5). By way of the duality between elements of $\partial p(u, v)$ and 'subderivatives' (see Sections 2–3), we thereby provide for the first time a general interpretation for such multiplier vectors. We also open the route to applying to p various fundamental theorems known about subgradients and subderivatives and we obtain in particular criteria for Lipschitz continuity that go well beyond previous ones. As a by-product, we get a new proof of Clarke's multiplier rule that shows it is valid under somewhat weaker assumptions, and also in a somewhat sharper form, than Clarke's or the version developed by Hiriart-Urruty [18] (see Section 4). We demonstrate that the known bounds on Dini derivatives of p follow from our subgradient estimates, without the restrictions on $(P_{u,v})$ that have been made in the past, and hold actually for Hadamard derivatives (see Section 7). We prove an extension of Golshtein's theorem for the mixed smooth-convex case of $(P_{u,v})$ that requires neither the set of optimal solutions nor the set of multiplier vectors to be compact.

A novel feature of our approach is that no form of implicit function theorem is ever used. At the critical stage we rely instead on our augmented Lagrangian results in [29].

2. Subderivatives and subgradients

The kind of subdifferential analysis initiated by Clarke for nonsmooth, non-convex functions has in the last several years been expanded and solidified in many ways. The lecture notes [30] can serve as an introduction to the finite-dimensional case with references. There is much to the subject that cannot be told here, but to assist the reader we shall touch on some of the central facts and definitions and do so in the notation of the function p. This will facilitate the applications we wish to make, although for the time being nothing dependent on the special nature of p as an optimal value function will be invoked.

Recall that a function $p : \mathbb{R}^m + \mathbb{R}^d \to (\mathbb{R} \cup \{\pm\infty\}$ is everywhere lower semicontinuous if and only if its epigraph

$$E = \{(u, v, \alpha) \in \mathbb{R}^m \times \mathbb{R}^d \times \mathbb{R} \mid \alpha \geq p(u, v)\} \tag{2.1}$$

is a closed set. In this case the matters we must explain are simpler, but we do not want to be burdened later with having to impose conditions on $(P_{u,v})$ that imply such *global* lower semicontinuity of its optimal value function. For our purposes all that really is needed is for the epigraph E to be closed relative to some neighborhood (in $\mathbb{R}^m \times \mathbb{R}^d \times \mathbb{R}$) of one of its points $(u, v, p(u, v))$ that happens to be under discussion. This condition is stronger than lower semicontinuity of p just at (u, v), yet not as stringent as requiring lower semicontinuity of p on a neighborhood of (u, v) (in $\mathbb{R}^m \times \mathbb{R}^d$). We shall call it *strict lower semicontinuity of p at* (u, v); it holds if and only if for some $\alpha > p(u, v)$, there is a neighborhood of (u, v) on which the function $\min\{p, \alpha\}$ is lower semicontinuous.

The epigraph point of view and the potential discontinuities of p also force us to be more subtle in speaking of convergence of (u', v') to (u, v). We introduce the notation

$$(u', v') \underset{p}{\to} (u, v) \Leftrightarrow \begin{cases} (u', v') \to (u, v), \\ p(u', v') \to p(u, v), \end{cases} \tag{2.2}$$

in situations where it is really just the convergence of the point $(u', v', p(u', v'))$ in E to $(u, v, p(u, v))$ that counts. Obviously, '$\to_p$' is the same as '$\to$' when p is continuous at (u, v), and in particular whenever p happens to be locally Lipschitzian.

We concentrate henceforth in this section and the next on a point (u, v) where p is finite and strictly lower semicontinuous. Criteria for this in the optimal value case will be given in Propositions 8–10 in Section 5.

Using the notation (2.2), we define the *Clarke derivative* of p at (u, v) with respect to a vector (h, k) as

$$p^\circ(u, v; h, k) = \limsup_{\substack{(u', v') \to_p (u, v) \\ t \downarrow 0}} \frac{p(u' + th, v' + tk) - p(u', v')}{t} \tag{2.3}$$

Clarke actually considered such derivatives only for locally Lipschitzian functions [4] ('$\to$' in place of '$\to_p$'), but he used them indirectly to develop a notion of 'subgradient' for functions that are merely lower semicontinuous and not necessarily finite-valued. We showed in [28] that the generalized subgradients in question could be characterized thoroughly and directly in terms of slightly more complicated limits than the ones in (2.3), namely the so-called *subderivatives*

$$p^{\uparrow}(u, v; h, k) = \lim_{\epsilon \downarrow 0} \limsup_{\substack{(u', v') \to_p (u, v) \\ t \downarrow 0}} \left[\inf_{\substack{|h'-h| \leq \epsilon \\ |k'-k| \leq \epsilon}} \left[\frac{p(u' + th', v' + tk') - p(u', v')}{t} \right] \right]. \tag{2.4}$$

The remarkable fact is that $p^{\uparrow}(u, v; h, k)$, as a function of (h, k), is always *convex*, positively homogeneous, lower semicontinuous, not identically $+\infty$ nor identically $-\infty$. Clarke's set of subgradients is given directly as

$$\partial p(u, v) = \{(y, z) \in \mathbb{R}^m \times \mathbb{R}^d \mid y \cdot h + z \cdot k \leq p^{\uparrow}(u, v; h, k) \text{ for all } (h, k)\}. \tag{2.5}$$

From this expression and the properties of the subderivative function it follows by general theorems of convex analysis [26, Section 13] that $\partial p(u, v)$ is a closed convex set and

$$\begin{aligned} p^{\uparrow}(u, v; h, k) &= \sup\{y \cdot h + z \cdot k \mid (y, z) \in \partial p(u, v)\} \\ &> -\infty \text{ for all } (h, k), \quad \text{if } \partial p(u, v) \neq \emptyset, \\ p^{\uparrow}(u, v; h, k) &= \pm\infty \text{ for all } (h, k) \quad \text{if } \partial p(u, v) = \emptyset. \end{aligned} \tag{2.6}$$

This formula extends one given by Clarke [4] for his derivatives (2.3) in the locally Lipschitzian case. In that case, $\partial p(u, v)$ is nonempty and compact; conversely, as we proved in [25], if $\partial p(u, v)$ is nonempty and compact, then p is locally Lipschitzian around (u, v) and the derivatives (2.3) coincide. A more general relationship between the two derivatives, established in [28, p. 267], is the following: the two effective domains

$$\text{dom } p^{\circ}(u, v; h, k) = \{(h, k) \mid p^{\circ}(u, v; h, k) < \infty\}, \tag{2.7}$$

$$\text{dom } p^{\uparrow}(u, v; h, k) = \{(h, k) \mid p^{\uparrow}(u, v; h, k) < \infty\}, \tag{2.8}$$

are convex cones containing $(0, 0)$ which have the same interior, and for (h, k) in this interior one has

$$\begin{aligned} \infty > \lim_{\epsilon \downarrow 0} &\left[\limsup_{\substack{(u', v') \to_p (u, v) \\ t \downarrow 0}} \left[\sup_{\substack{|h'-h| \leq \epsilon \\ |k'-k| \leq \epsilon}} \frac{p(u' + th', v' + tk') - p(u', v')}{t} \right] \right] \\ &= p^{\uparrow}(u, v; h, k) = p^{\circ}(u, v; h, k). \end{aligned} \tag{2.9}$$

With respect to such vectors (h, k), p is said to be *directionally Lipschitzian*. (This concept generalizes Lipschitz continuity in a neighborhood of (u, v), which is the case of $(h, k) = (0, 0)$; then the cones in (2.7) and (2.8) are the whole space, and (2.9) actually holds for all (h, k), with '$\to_p$' identical to '$\to$'.) For the many consequences and uses of the directionally Lipschitzian property, see [28, 27].

Formulas (2.5) and (2.6) underline the complete duality between subderivatives and subgradients. If p is convex, $\partial p(u, v)$ is identical to the subgradient set of convex analysis, while if p is smooth it reduces to the singleton

$\{\nabla p(u, v)\}$ [4, 28]. Indeed, $\partial p(u, v)$ consists of a single vector (y, z) if and only if $p^{\uparrow}(u, v; h, k)$, or equivalently $p^{\circ}(u, v; h, k)$, is linear in (h, k), and in this event p is *strictly* differentiable at (u, v) with $\nabla p(u, v) = (y, z)$:

$$\lim_{\substack{(h',k')\to(h,k)\\(u',v')\to(u,v)\\t\downarrow 0}} \frac{p(u'+th', v'+tk') - p(u', v')}{t} = y \cdot h + z \cdot k \tag{2.10}$$

[4, 28]. The implication of this result for our later efforts, incidentally, is that differentiability of p at (u, v) can be deduced from conditions which imply $\partial p(u, v)$ has exactly one element.

In general, bounds on various derivatives of p can be obtained from estimates for $\partial p(u, v)$, and this is the pattern we shall follow. Besides $p^{\circ}(u, v; h, k)$ and $p^{\uparrow}(u, v; h, k)$ we shall consider upper and lower one-sided *Hadamard* derivatives:

$$p^{+}(u, v; h, k) = \limsup_{\substack{(h',k')\to(h,k)\\t\downarrow 0}} \frac{p(u+th', v+tk') - p(u, v)}{t}, \tag{2.11}$$

$$p_{+}(u, v; h, k) = \liminf_{\substack{(h',k')\to(h,k)\\t\downarrow 0}} \frac{p(u+th', v+tk') - p(u, v)}{t}. \tag{2.12}$$

Obviously one always has

$$p_{+}(u, v; h, k) \le p^{\uparrow}(u, v; h, k), \tag{2.13}$$

and if p is directionally Lipschitzian at (u, v) with respect to (h, k), so that (2.9) holds, then also

$$p^{+}(u, v; h, k) \le p^{\circ}(u, v; h, k). \tag{2.14}$$

The case where equality holds in (2.13) plays an important role in the literature; then we say p is *subdifferentially regular* at (u, v) (cf. [4, 28, 25]).

Note that when $p^{+}(u, v; h, k) = p_{+}(u, v; h, k)$ one has a property stronger than just the existence of $p'(u, v; h, k)$ as defined in (1.1). This is what we will be able to establish in Section 7 in cases where other authors have considered only $p'(u, v; h, k)$, as well as made other restrictions.

In some situations it is crucial to be able to know at least that $\partial p(u, v)$ is nonempty. As recorded already in (2.6), a necessary and sufficient condition for this is the existence of (h, k) such that $p^{\uparrow}(u, v; h, k)$ is finite. We now elaborate the meaning of this.

Proposition 1. *Under the assumption that p is finite and lower semicontinuous at (u, v), one has $\partial p(u, v) \neq \emptyset$ if and only if there exist sequences $t_j \downarrow 0$ and $(u^j, v^j) \to_p (u, v)$ such that for no convergent sequence $(h^j, k^j) \to (h, k)$ does one have*

$$[p(u^j + t_j h^j, v^j + t_j k^j) - p(u^j, v^j)]/t_j \to -\infty.$$

Thus in particular, $\partial p(u, v) \neq \emptyset$ *if* p *is calm at* (u, v) *in the sense that*

$$\liminf_{(u', v') \to (u, v)} \frac{p(u', v') - p(u, v)}{|(u', v') - (u, v)|} > -\infty. \tag{2.15}$$

Proof. Because the function $p^{\uparrow}(u, v; \cdot, \cdot)$ is lower semicontinuous, positively homogeneous and convex, but not identically $+\infty$, it is finite at some point if and only if it is not $-\infty$ at the origin. Therefore, $\partial p(u, v) \neq \emptyset$ if and only if $p^{\uparrow}(u, v; 0, 0) > -\infty$. The first assertion in the proposition merely puts the latter condition in more specific terms using the definition (2.4). The calmness property implies the condition is satisfied with $(u^j, v^j) = (u, v)$ for all j and any sequence $t_j \downarrow 0$.

Calmness of p at (u, v) may be thought of as 'pointwise lower Lipschitz continuity'. It is a concept that has been used to advantage by Clarke in [3].

3. Singular subgradients

In addition to the subgradients discussed so far, we shall find it helpful to speak of as *singular subgradients* of p at (u, v) the elements of the closed convex cone

$$\begin{aligned} \partial^0 p(u, v) :&= \text{polar of the convex cone (2.8)} \\ &= \{(y, z) \mid y \cdot h + z \cdot k \leq 0 \text{ for all } (h, k) \text{ satisfying} \\ &\quad\ p^{\uparrow}(u, v; h, k) < \infty\}. \end{aligned} \tag{3.1}$$

It follows from the duality in (2.5) and (2.6) that this set is just the recession cone of $\partial p(u, v)$ [26, Section 13]:

$$\partial^0 p(u, v) = 0^+ \partial p(u, v) = \limsup_{\lambda \downarrow 0} \lambda \partial p(u, v) \quad \text{when } \partial p(u, v) \neq \emptyset. \tag{3.2}$$

Nonzero singular subgradients thus describe directions which can be identified with 'elements of $\partial p(u, v)$ lying at ∞', except that there can be situations where $\partial p(u, v) = \emptyset$ and yet $\partial^0 p(u, v) \neq \emptyset$.

A more geometric description of singular subgradients is possible in terms of Clarke's concept of normal cones to closed sets in Euclidean spaces. Recall from the beginning of Section 2 that when p is finite and *strictly* lower semicontinuous at (u, v), its epigraph E is closed relative to a neighborhood of the point $(u, v, p(u, v))$. The *normal cone* to E at this point is the nonempty closed convex cone

$$N_E(u, v, p(u, v)) = \partial \delta_E(u, v, p(u, v)), \tag{3.3}$$

where δ_E is the indicator function for E. One has

$$\partial p(u, v) = \{(y, z) \mid (y, z, -1) \in N_E(u, v, p(u, v))\}, \tag{3.4}$$

$$\partial^0 p(u, v) = \{(y, z) \mid (y, z, 0) \in N_E(u, v, p(u, v))\}. \tag{3.5}$$

In Clarke's original approach [4], normal cones are first given various direct characterizations, and then (3.4) is taken as the *definition* of the set of subgradients of p at (u, v). As seen from (3.5), the notion of 'singular subgradients' fits neatly into the same picture. The validity of (3.5) stems from the fact that the cone $N_E(u, v, p(u, v))$ and the epigraph of the subderivative function $p^{\uparrow}(u, v; \cdot, \cdot)$ are polar to each other; see [28].

(Incidentally, the assertion made in Section that $p^{\uparrow}(u, v; h, k)$ cannot be identically $-\infty$ as a function of (h, k) follows by duality from the fact that $N_E(u, v, p(u, v))$ cannot consist of just the zero vector. The latter is true because $(u, v, p(u, v))$ is a boundary point of E, and nonzero normal vector always exist at boundary points [25, p. 149].)

Several properties of p can be characterized in terms of singular subgradients, and this will be useful later in seeing the consequences of the estimates that will be given for $\partial^0 p(u, v)$. The following terminology will expedite matters: a cone M (not necessarily convex) will be called *pointed* if the origin cannot be expressed as a sum of nonzero vectors in M. When M is convex (as in the case $M = \partial^0 p(u, v)$), this reduces to the property that M does not contain the negative of any of its nonzero vectors.

Proposition 2. *Under the assumption that p is finite and strictly lower semicontinuous at (u, v), one has p directionally Lipschitzian with respect to (h, k) if and only if for all (h', k') in some neighborhood of (h, k), one has $y \cdot h' + z \cdot k' \le 0$ for all $(y, z) \in \partial^0 p(u, v)$. Such an (h, k) exists if and only if $\partial^0 p(u, v)$ is pointed.*

Proof. The condition says that (h, k) is an interior point of the polar of $\partial^0 p(u, v)$. Since $\partial^0 p(u, v)$ is the polar of the convex cone (2.8), this means (h, k) belongs to the interior of (2.8). Such vectors (h, k) are the ones with respect to which p is directionally Lipschitzian, as already explained in Section 2. The polar of a closed convex cone has nonempty interior if and only if the cone is pointed.

Proposition 3. *For p to be locally Lipschitzian around (u, v), it is necessary and sufficient that p be finite and strictly lower semicontinuous at (u, v) and have $\partial^0 p(u, v) = \{(0, 0)\}$.*

Proof. This is the case of Proposition 2 where $(h, k) = (0, 0)$. Recall that p is locally Lipschitzian around (u, v) if and only if p is directionally Lipschitzian at (u, v) with respect to $(h, k) = (0, 0)$ [28].

Proposition 4. *Under the assumption that p is finite and strictly lower semicontinuous at (u, v), if $\partial^0 p(u, v)$ is pointed and does not contain any vector of the form $(y, 0)$ with $y \neq 0$, then*

$$\partial_v p(u, v) \subset \{z \mid \exists y \text{ with } (y, z) \in \partial p(u, v)\}, \tag{3.6}$$

$$\partial_v^0 p(u, v) \subset \{z \mid \exists y \text{ with } (y, z) \in \partial^0 p(u, v)\}. \tag{3.7}$$

In particular, (3.6) is valid if p is locally Lipschitzian around (u, v). (Moreover, equality holds in (3.6) and (3.7) if p is subdifferentially regular at (u, v).)

Proof. From a result in [27, p. 350], (3.6) holds (and with equality in the case of subdifferential regularity) when the interior of the convex cone (2.8) contains a vector of form $(0, k)$. The separation theorem for convex sets enables us to translate this condition into the nonexistence of a vector $(y, 0) \neq (0, 0)$ belonging to the polar of the cone (2.8), namely $\partial^0 p(u, v)$ (cf. Proposition 2). The locally Lipschitzian case of (3.6) follows via Proposition 3. There are several ways to get the parallel inclusion (3.7), but the simplest perhaps is to observe that the cited result in [27, p. 350] is a corollary of a theorem that actually yields more when specialized to the case in question: for the function $q = p(u, \cdot)$, one has

$$q^{\uparrow}(v; k) \leq p^{\uparrow}(u, v; 0, k) \quad \text{for all } k, \tag{3.8}$$

(and equality holds in (3.8) when p is subdifferentially regular at (u, v)). Therefore

$$\{(h, k) \mid h = 0,\ q^{\uparrow}(v; k) < \infty\} \supset \{(h, k) \mid p^{\uparrow}(u, v; h, k) < \infty\} \cap [\{0\} \times \mathbb{R}^d]. \tag{3.9}$$

Since the interior of the cone (2.8) contains under our hypothesis a vector of form $(0, k)$, we can take polars on both sides of (3.9) and get

$$[\mathbb{R}^m \times \partial^0 q(v)] \subset \partial^0 p(u, v) + [\mathbb{R}^m \times \{0\}],$$

which is equivalent to (3.7).

Remark. The Lipschitzian case of Proposition 4 was first developed by Clarke, who pointed out that without some condition like subdifferential regularity, there may be no inclusion either way between $\partial p(u, v)$ and $\partial_u p(u, v) \times \partial_v p(u, v)$. See [17, p. 308] for an example of this phenomenon.

4. Lagrange multiplier rule

Our main result about subgradients of p when p is the optimal value function in Section 1 will involve Lagrange multiplier vectors that appear in extended first-order necessary conditions for optimality in $(P_{u,v})$. This section is devoted to formulating the conditions in question and comparing them to previous contributions. The necessity of the conditions, however, will actually be established in Section 6 as a *consequence* of our estimation theorem, rather than as a preliminary to it.

Henceforth our notation and assumptions are those in Section 1, but we apply freely the general subdifferential theory exposed in Sections 2–3.

Each function f_i, being locally Lipschitzian on an open set containing D, has a nonempty, compact, convex subgradient set $\partial f_i(v, x)$ at every $(v, x) \in D$. We emphasize that this is the subgradient set of convex analysis if f_i is a convex function, and it is just $\{\nabla f_i(v, x)\}$ if f_i is of class $\mathscr{C}^1$. When f_i is mixed smooth-convex as in case (c) of Section 1, it turns out that

$$\partial f_i(v, x) = (\nabla_v f_i(v, x), \partial_x f_i(v, x)) \tag{4.1}$$

(because $f_i^\circ = f_i'$ in this case, as can be verified by direct calculation).

Since D is closed, the indicator function

$$\delta_D(v, x) = \begin{cases} 0, & \text{if } (v, x) \in D, \\ \infty, & \text{if } (v, x) \notin D, \end{cases}$$

is lower semicontinuous everywhere. Its subgradient sets are the normal cones to D:

$$N_D(v, x) = \partial\delta_D(v, x) \quad \text{for each } (v, x) \in D. \tag{4.2}$$

(When D is convex, the vectors $(z, w) \in N_D(v, x)$ are the ones such that $(z, w) \cdot (v', x') \leq (z, w) \cdot (v, x)$ for all $(v', x') \in D$.)

The optimality conditions we shall be concerned with are related to such subgradients, as will be explained below, but they generally take the form of associating with some x which in particular satisfies all the constraints of $(\text{P}_{u,v})$ a pair of vectors $y = (y_1, \ldots, y_m)$ and $z = (z_1, \ldots, z_d)$ such that

$$y_i \geq 0 \quad \text{and} \quad y_i[f_i(v, x) + u_i] = 0 \quad \text{for } i = 1, \ldots, s, \tag{4.3}$$

$$(z, 0) \in \partial\left[f_0 + \sum_{i=1}^m y_i f_i + \delta_D\right](v, x). \tag{4.4}$$

For some purposes, we shall need to look at the corresponding degenerate conditions where f_0 does not appear, i.e., where (4.4) is replaced by

$$(z, 0) \in \partial\left[\sum_{i=j}^m y_i f_i + \delta_D\right](v, x). \tag{4.5}$$

We let

$$K(u, v, x) = \text{set of all } (y, z) \text{ satisfying (4.3) and (4.4)},$$

$$K_0(u, v, x) = \text{set of all } (y, z) \text{ satisfying (4.3) and (4.5)}. \tag{4.6}$$

The targeted Lagrange multiplier rule is an assertion that $K(u, v, x) \neq \emptyset$ in certain situations. For immediate comparison with classical conditions, observe that in the smooth case (a) of Section 1, (4.4) reduces to

$$0 = \nabla_x f_0(v, x) + \sum_{i=1}^m y_i \nabla_x f_i(v, x) \quad \text{and} \quad z = \nabla_v f_0(v, x) + \sum_{i=1}^m y_i \nabla_v f_i(v, x), \tag{4.7}$$

while (4.5) reduces to

$$0 = \sum_{i=1}^m y_i \nabla_x f_i(v, x) \quad \text{and} \quad z = \sum_{i=1}^m y_i \nabla_v f_i(v, x). \tag{4.8}$$

Since $(0, 0) \in K_0(u, v, x)$ trivially always, interest in the set $K_0(u, v, x)$ will center on whether it also contains some $(y, z) \neq (0, 0)$. The condition $K_0(u, v, x) = \{(0, 0)\}$ will serve as one kind of constraint qualification. A more subtle constraint qualification that will also play a role can be stated in terms of 'calmness'.

Localizing a definition of Clarke's [3], we say problem $(P_{u,v})$ is *calm at x*, one of its locally optimal solutions, if there do *not* exist sequences $x^j \to x$ and $(u^j, v^j) \to (u, v)$ with x^j feasible for (P_{u^j,v^j}) such that

$$\frac{f_0(v^j, x^j) - f_0(v, x)}{|(u^j, v^j) - (u, v)|} \to -\infty.$$

Clearly this does hold when p is calm at (u, v) in the sense of (2.15) and x is any (globally) optimal solution to $(P_{u,v})$. Calmness of p at (u, v), without reference additionally to any point x, is a condition that Clarke calls simply the calmness of problem $(P_{u,v})$. The exact relationship between this 'global' calmness and our 'local' calmness will be shown later in Proposition 12 (see Section 6).

Theorem 1. *Let x be any locally optimal solution to $(P_{u,v})$.*
(i) *If $(P_{u,v})$ is calm at x, then $K(u, v, x) \neq \emptyset$.*
(ii) *If $K_0(u, v, x) = \{(0, 0)\}$, then $(P_{u,v})$ is indeed calm at x, and moreover $K(u, v, x)$ is compact.*

As already remarked, this theorem will not be proved until Section 6, where it will appear chiefly as a sort of corollary of Theorem 2 of Section 5. We have stated it at this early stage in order to put the multiplier sets $K(u, v, x)$ and $K_0(u, v, x)$ in the proper perspective. The rest of this section deals with further clarifications of the nature of these sets. We start by citing a fundamental rule of subdifferential calculus.

Proposition 5. *Let g_1 and g_2 be extended-real-valued functions on a Euclidean space which are both finite at a point w. If either g_1 or g_2 is locally Lipschitzian around w, then*

$$\partial(g_1 + g_2)(w) \subset \partial g_1(w) + \partial g_2(w).$$

Moreover, equality holds if either g_1 or g_2 is of class $\mathscr{C}^1$ in a neighborhood of w, or if both g_1 and g_2 are subdifferentially regular at w.

Proof. This is an immediate consequence of a much broader result obtained in [27, p. 345], except for the business about g_1 or g_2 being of class $\mathscr{C}^1$. If g_2, say, is of class $\mathscr{C}^1$ around w, then g_2 and $-g_2$ are both locally Lipschitzian around w and have $\partial g_2(w) = \{\nabla g_2(w)\}$ and $\partial(-g_2)(w) = \{-\nabla g_2(w)\}$. The basic rule gives both

$$\partial(g_1 + g_2)(w) \subset \partial g_1(w) + \nabla g_2(w)$$

and

$$\partial g_1(w) = \partial(g_1 + g_2 - g_2)(w) \subset \partial(g_1 + g_2)(w) - \nabla g_2(w),$$

and this implies $\partial(g_1 + g_2)(w) = \partial g_1(w) + \nabla g_2(w)$ and finishes the proof.

In the situation at hand, we want to apply Proposition 5 to the expressions in (4.4) and (4.5) along with the elementary rule that (inasmuch as f_i is locally Lipschitzian)

$$\partial(y_i f_i)(v, x) = y_i \partial f_i(v, x) \quad \text{for all } y_i \in \mathbb{R}. \tag{4.9}$$

For this purpose we note that the property of *subdifferential regularity* (see Section 2) holds everywhere for f_i when f_i is convex, of class $\mathscr{C}^1$, or a mixture of the two as in case (c) in Section 1. It holds everywhere for both f_i and $-f_i$ (i.e., for $y_i f_i$ regardless of the sign of y_i) if and only if f_i is of class $\mathscr{C}^1$. It holds for δ_D if and only if D is *tangentially regular* in the sense that at all boundary points of D, the Clarke tangent cone and the classical contingent cone coincide, as is true certainly when D is convex or a 'smooth manifold'; see [4, 27] for more on such properties.

At all events, the strong form of Proposition 5, where equality holds, is thoroughly applicable (together with (4.9)) in cases (a), (b) and (c) of Section 1 and more generally in the following cases of problem $(\mathrm{P}_{u,v})$:

(d) *the subdifferentially regular case*: D tangentially regular, f_1 subdifferentially regular for $i = 0, 1, \ldots, s$ and of class $\mathscr{C}^1$ for $i = s+1, \ldots, m$.

(e) *the extended smooth case*: D an arbitrary closed set, every f_i of class $\mathscr{C}^1$.

Clearly (e) subsumes (a), while from the remarks above, (d) subsumes (a), (b) and (c). This allows us to draw an important conclusion.

Proposition 6. *In condition* (4.4) *of the definition of* $K(u, v, x)$, *one has*

$$\partial\left[f_0 + \sum_{i=1}^m y_i f_i + \delta_D\right](v, x) \subset \partial\left[f_0 + \sum_{i=1}^m y_i f_i\right](v, x) + N_D(v, x)$$

$$\subset \partial f_0(v, x) + \sum_{i=1}^m y_i \partial f_i(v, x) + N_D(v, x). \tag{4.10}$$

Moreover, equality holds in cases (d) *and* (e) *above and hence in particular in the smooth, convex, and mixed smooth-convex cases* (a), (b) *and* (c) *of* $(\mathrm{P}_{u,v})$. *Similarly for condition* (4.5) *of the definition of* $K_0(u, v, x)$.

The second inclusion in Proposition 6 does not depend on the full force of Proposition 5: it is already apparent from an earlier formula of Clarke [4] where g_1 and g_2 are *both* locally Lipschitzian.

Observe that in the mixed smooth-convex case (c), where (4.1) holds and $D = \mathbb{R}^d \times C$, Proposition 6 allows conditions (4.4) and (4.5) to be written instead as

$$0 \in \partial_x\left[f_0 + \sum_{i=1}^m y_i f_i\right](v, x) + N_C(x) \quad \text{and } z = \nabla_v f_0(v, x) + \sum_{i=1}^m y_i \nabla_v f_i(v, x), \tag{4.11}$$

$$0 \in \partial_x\left[\sum_{i=1}^m y_i f_i\right](v, x) + N_C(x) \quad \text{and} \quad z = \sum_{i=1}^m y_i \nabla_v f_i(v, x). \tag{4.12}$$

Due to convexity in x, these indicate that when $(y, z) \in K(u, v, x)$, the pair (x, y) is (as expected) a saddlepoint of the ordinary Lagrangian for $(P_{u,v})$ on $C \times [\mathbb{R}^s_+ \times \mathbb{R}^{m-s}]$, and similarly when $(y, z) \in K_0(u, v, x)$, except that then it is the degenerate Lagrangian not involving f_0.

Only in situations where strict inclusions can be encountered in (4.10), and thus never in cases (a), (b), (c), (d) or (e), is the multiplier condition $K(u, v, x) \neq \emptyset$ in Theorem 1 any sharper than the ones of Clarke [3] or Hiriart-Urruty [18]. Clarke's rule corresponds to substituting the largest of the sets in (4.10) for (4.4), while Hiriart-Urruty uses the middle set.

These earlier rules do not actually take the parameter vector v into account, but they can be adapted to yield conditions in the present format simply by posing $(P_{u,v})$ equivalently as the problem $(P'_{u,v})$ in Section 1. Conversely, of course, Theorem 1 can be applied with v held fixed and suppressed from consideration. The corresponding multiplier conditions then say nothing about a vector z, and they have in place of (4.4) and (4.5) the relations

$$0 \in \partial_x \left[f_0 + \sum_{i=1}^{m} y_i f_i + \delta_D \right](v, x). \tag{4.13}$$

$$0 \in \partial_x \left[\sum_{i=1}^{m} y_i f_i + \delta_D \right](v, x), \tag{4.14}$$

which again could be elaborated as in Proposition 6. As far as necessary conditions for optimality are concerned, there is no distinction to be made between the two formulations in cases (a) or (c) (where (4.4), (4.5), become (4.7), (4.8), or (4.11), (4.12)). Nor is there any real distinction in the convex case (b), or for that matter in the subdifferentially regular case (d): then (4.13) holds if and only if there exists z such that (4.4) holds (apply the equality clause in Proposition 4 to the functions in question). Generally speaking, however, neither formulation of the conditions directly subsumes the other.

In the smooth case (a), the constraint qualification $K_0(u, v, x) = \{(0, 0)\}$ asserts:

$$\text{there is no } y \neq 0 \text{ satisfying (4.3) with } \sum_{i=1}^{m} y_i \nabla_x f_i(v, x) = 0. \tag{4.15}$$

This property is equivalent by duality with the *Mangasarian–Fromovitz constraint qualification* [24]:

the gradients $\nabla_x f_i(v, x)$, $i = s+1, \ldots, m$, are linearly independent, and there is a vector w such that

$$\nabla_x f_i(v, x) \cdot w \begin{cases} < 0 & \text{for } i = 1, \ldots, s \text{ having } f_i(v, x) = 0, \\ = 0 & \text{for } i = s+1, \ldots, m. \end{cases} \tag{4.16}$$

Related conditions for nonsmooth cases of $(P_{u,v})$ have been introduced by Auslender [2] and Hiriart-Urruty [18, 19]. Our condition $K_0(u, v, x) = \{(0, 0)\}$ is sharper than these in the sense of the inclusions in Proposition 6, but Hiriart-Urruty gives a treatment of equality constraints that is in other respects more refined. On the other hand, Hiriart-Urruty does not prove a multiplier rule based on 'calmness'.

The result in Theorem 1 that the constraint qualification $K_0(u, v, x) = \{(0, 0)\}$ implies calmness at x is new, although in the extended smooth case (e) with D convex it follows in terms of the Mangasarian–Fromovitz qualification via the stability theory of Robinson [31, 32]; c.f. remark of Clarke [3, p. 173].

There is a relationship between $K_0(u, v, x)$ and $K(u, v, x)$ that sheds some further light. Recall that the *recession cone* of the (not necessarily convex) set $K(u, v, x)$ is by definition

$$
\begin{aligned}
0^+K(u, v, x) &= \limsup_{\lambda \downarrow 0} \lambda K(u, v, x) \\
&= \{\lim \lambda_j(y^j, z^j) \mid \lambda_j \downarrow 0, (y^j, z^j) \in K(u, v, x)\}. \qquad (4.17)
\end{aligned}
$$

A nonempty set in a Euclidean space is bounded if and only if its recession cone consists of just the zero vector.

Proposition 7. *For any feasible solution x to* ($P_{u,v}$), *the sets $K(u, v, x)$ and $K_0(u, v, x)$ are closed and*

$$0^+K(u, v, x) \subset K_0(u, v, x). \qquad (4.18)$$

In cases (d) *and* (e) *above* (*and hence in particular in the smooth, convex, and mixed smooth-convex cases* (a), (b) *and* (c), $K(u, v, x)$ *and* $K_0(u, v, x)$ *are also convex. If in addition to this $K(u, v, x)$ is nonempty, then equality holds in* (4.18) *and*

$$K(u, v, x) + K_0(u, v, x) = K(u, v, x). \qquad (4.19)$$

Proof. To demonstrate that $K(u, v, x)$ is closed, suppose $(y^j, z^j) \in K(u, v, x)$ and $(y^j, z^j) \to (y, z)$. For all j, one has

$$y_i^j \begin{cases} \geq 0 & \text{for } i = 1, \ldots, s \text{ having } f_i(v, x) = 0, \\ = 0 & \text{for } i = 1, \ldots, s \text{ having } f_i(v, x) < 0, \end{cases}$$

so the same holds for the multipliers $y_i = \lim_j y_i^j$. Also

$$
\begin{aligned}
(z^j, 0) &\in \partial\left[f_0 + \sum_{i=1}^m y_i^j f_i + \delta_D\right](v, x) \\
&= \partial\left[f_0 + \sum_{i=1}^m y_i f_i + \delta_D + \sum_{i=1}^m (y_i^j - y_i) f_i\right](v, x). \qquad (4.20)
\end{aligned}
$$

Applying Proposition 5, we get

$$(z^j, 0) \in \partial\left[f_0 + \sum_{i=1}^m y_i f_i + \delta_D\right](v, x) + \sum_{i=1}^m (y_i^j - y_i)\partial f_i(v, x).$$

Since $z^j \to z$, $y_i^j - y_i \to 0$, and $\partial f_i(v, x)$ is compact (due to f_i being locally Lipschitzian), it follows that (4.4) holds. Thus $(y, z) \in K(u, v, x)$, and $K(u, v, x)$ is closed. The proof that $K_0(u, v, x)$ is closed is identical.

The proof of the inclusion (4.18) is along similar lines. Suppose $\lambda_j(y^j, z^j) \to (y, z)$, where $\lambda_j \downarrow 0$ and $(y^j, z^j) \in K(u, v, x)$. The critical observation this time is that (4.20) can be written instead in the form

$$\begin{aligned}(\lambda_j z^j, 0) &\in \partial\Big(\lambda_j\Big[f_0 + \sum_{i=1}^m y_i^j f_i + \delta_D\Big]\Big)(v, x).\\ &= \partial\Big[\sum_{i=1}^m y_i f_i + \delta_D + \lambda_j f_0 + \sum_{i=1}^m (\lambda_j y_i^j - y_i) f_i\Big](v, x),\end{aligned}$$

so that by Proposition 5

$$(\lambda_j z^j, 0) \in \partial\Big[\sum_{i=1}^m y_i f_i + \delta_D\Big](v, x) + \lambda_j \partial f_0(v, x) + \sum_{i=1}^m (\lambda_j y_i^j - y_i)\partial f_i(v, x).$$

Since $\lambda_j z^j \to z$, $\lambda_j y_i^j - y_i \to 0$, $\lambda_j \downarrow 0$ and $\partial f_i(v, x)$ is compact, we get (4.5) in the limit and hence $(y, z) \in K_0(u, v, x)$.

In cases (d) and (e), we know that equality holds in (4.10) and that $\partial f_i(v, x)$ is just a singleton for $i = s+1, \ldots, m$. Using this in (4.4), it is easy to verify the convexity of $K(u, v, x)$ and similarly that of $K_0(u, v, x)$, as well as the relation

$$K(u, v, x) + K_0(u, v, x) \subset K(u, v, x). \tag{4.21}$$

(Recall that $(\alpha + \beta)C = \alpha C + \beta C$ when C is a nonempty convex set and $\alpha \geq 0$, $\beta \geq 0$; cf. [26, Section 3].) When $K(u, v, x)$ is convex and nonempty, (4.21) implies $K_0(u, v, x) \subset 0^+K(u, v, x)$ [26, Section 8], whence equality in (4.18) and (4.19).

Remark. In the convex case (b), the condition $K(u, v, x) \neq \emptyset$ is, of course, sufficient for a feasible solution x in $(P_{u,v})$ to be optimal. Indeed, the multiplier relations reduce then to the description of a saddlepoint of the Lagrangian for the equivalent problem $(P'_{u,v})$ in Section 1. Because of this, the set $K(u, v, x)$ is actually the same regardless of which optimal solution x is being considered, and similarly for $K_0(u, v, x)$. Another special result in the convex case, besides the ones noted in Propositions 6 and 7, is the converse of Theorem 1(i): if $K(u, v, x) \neq \emptyset$ for an optimal solution x, then $(P_{u,v})$ is calm at x; in fact p is calm at (u, v). For this, see [26, Sections 28–29].

5. Tameness and subgradient estimates

Our main result will be stated in this section after some preliminaries having to do with lower semicontinuity of the optimal value function p and the existence of solutions to $(P_{u,v})$.

We shall say for a given (u, v) that problem $(P_{u,v})$ is *tame* if there is a set

$A \subset \mathbb{R}^n$ with the property:

$$\begin{aligned}&A \text{ is compact, and for every}\\ &\epsilon > 0 \text{ there exist } \delta > 0 \text{ and } \alpha > p(u, v) \text{ such that}\\ &\text{when } |(u', v') - (u, v)| < \delta \text{ and } p(u', v') < \alpha, \text{ the addition of the}\\ &\text{constraint } \operatorname{dist}(x, A) \le \epsilon \text{ to } (\mathrm{P}_{u', v'}) \text{ would not}\\ &\text{affect the infimum } p(u', v') \text{ in } (\mathrm{P}_{u', v'}).\end{aligned} \tag{5.1}$$

The virtues of this condition are proclaimed in the next three propositions. (Recall the meaning of 'strict' lower semicontinuity, as defined at the beginning of Section 2.) Note that 'tameness' is *not* a constraint qualification like 'calmness', but merely a weak sort of local boundedness assumption on the way the feasible solution set varies with the parameters.

Proposition 8. *Suppose (u, v) is such that $(\mathrm{P}_{u,v})$ is tame in the above sense. Then p is finite at (u, v) and strictly lower semicontinuous at (u, v). Furthermore $(\mathrm{P}_{u,v})$ has at least one optimal solution; indeed, if A is any set with respect to which the definition of tameness is fulfilled, then $(\mathrm{P}_{u,v})$ must have an optimal solution lying in A.*

Proof. Taking $(u', v') = (u, v)$ in (5.1), we see in particular that $p(u, v) < \infty$. Define

$$\beta = \liminf_{(u', v') \to (u, v)} p(u', v') \le p(u, v). \tag{5.2}$$

Select any sequence $\epsilon_j \downarrow 0$ and corresponding sequences of numbers δ_j and α_j. In view of (5.2), a sequence $(u^j, v^j) \to (u, v)$ with $p(u^j, v^j) \to \beta$ exists having actually $|(u^j, v^j) - (u, v)| < \delta_j$ and $p(u^j, v^j) < \alpha_j$. Then for each j, (P_{u^j, v^j}) has feasible solutions which also belong to the set

$$\{x \mid \operatorname{dist}(x, A) \le \epsilon_j\} \tag{5.3}$$

(which is compact because A is compact), and the infimum is unaffected if restricted to such feasible solutions. Since the objective function in (P_{u^j, v^j}) is continuous and the set of all feasible solutions is closed, it follows that (P_{u^j, v^j}) has an optimal solution x^j in the set (5.3). Then $f_0(v^j, x^j) = p(u^j, v^j) \to \beta$ and $\operatorname{dist}(x^j, A) \to 0$. Passing to subsequences if necessary, we can suppose (again because A is compact) that $x^j \to x$, where x is some element of A. The continuity of the functions f_i and the closedness of the set D imply that, since $(u^j, v^j) \to (u, v)$, x is a feasible solution to $(\mathrm{P}_{u,v})$ with $f_0(v, x) = \beta$. We may conclude then from (5.2) that x is optimal and $\beta = p(u, v)$.

Proposition 9. *A sufficient condition for (u, v) to be such that $(\mathrm{P}_{u,v})$ is tame is the existence of $\delta_0 > 0$ and $\alpha_0 > p(u, v)$ with the property: the set of all x' satisfying*

$$\begin{aligned}&\exists (u', v') \text{ with } |(u', v') - (u, v)| \le \delta_0 \text{ such that}\\ &x' \text{ is feasible for } (\mathrm{P}_{u', v'}) \text{ and } f_0(v', x') \le \alpha_0\end{aligned} \tag{5.4}$$

is a bounded set. Indeed, the definition of tameness is then fulfilled with this set as A.

In particular, $(\mathrm{P}_{u,v})$ *is tame if it has feasible solutions and D is of the form* $\mathbb{R}^d \times C$ *with C compact. (Then C can serve as the A in the definition of tameness.)*

Proof. Denote the set of x' satisfying (5.4) by A and observe that it is compact. To verify the rest of (5.1) consider any $\epsilon > 0$ and let $\delta = \delta_0$, $\alpha = \alpha_0$. Then for (u', v') with $|(u', v')| - (u, v)| < \delta$ and $p(u', v') < \alpha$, all the feasible solutions x' to $(\mathrm{P}_{u',v'})$ with $f_0(v', x') \le \alpha$ belong to A and therefore satisfy $\operatorname{dist}(x', A) = 0$. Hence the constraint $\operatorname{dist}(x, A) \le \epsilon$ can be added to $(\mathrm{P}_{u',v'})$ with impunity.

Proposition 10. *A necessary and sufficient condition for* (u, v) *to be such that* $(\mathrm{P}_{u,v})$ *is tame is the existence of* $\delta_0 > 0$ *and* $\alpha_0 > p(u, v)$ *with the property: there is a bounded mapping* ξ *from the set*

$$\{(u', v') \mid p(u', v') < \alpha_0 \quad \text{and} \quad |(u', v') - (u, v)| < \delta_0\} \tag{5.5}$$

to $\mathbb{R}^n$ *such that for every* (u', v') *in this set,* $\xi(u', v')$ *is an optimal solution to* $(\mathrm{P}_{u',v'})$.

Indeed, the definition of tameness is satisfied with respect to a particular compact set A if and only if for some such mapping ξ, *A includes all the cluster points of* $\xi(u', v')$ *as* $(u', v') \to_p (u, v)$ *in the sense of* (2.2). *(These cluster points themselves form a compact set of optimal solutions to* $(\mathrm{P}_{u,v})$.)

Proof. If there is such a mapping ξ, and C denotes its set of cluster points of $\xi(u', v')$ as $(u', v') \to_p (u, v)$, then C is a compact set of points x which (by the closedness of D and continuity of f_i) are feasible solutions to $(\mathrm{P}_{u,v})$ having $f_0(v, x) = p(u, v)$. Thus C consists of optimal solutions to $(\mathrm{P}_{u,v})$, and for any $\epsilon > 0$ there exist $\delta > 0$ and $\alpha > p(u, v)$ such that whenever (u', v') satisfies $|(u', v') - (u, v)| < \delta$ and $p(u', v') < \alpha$, one has $\operatorname{dist}(\xi(u', v'), C) \le \epsilon$. Since $\xi(u', v')$ is optimal for $(\mathrm{P}_{u',v'})$, it follows that for such (u', v') the constraint $\operatorname{dist}(x', C) \le \epsilon$ can be added to $(\mathrm{P}_{u',v'})$ without affecting the infimum in the problem. The same then holds for any compact $A \supset C$; such an A therefore satisfies (5.1).

Conversely, suppose A is a set with property (5.1). Choose any sequence $\epsilon_j \downarrow 0$ (starting with $j = 0$) and corresponding values δ_j and α_j as in (5.1); the latter values can systematically be lowered, if necessary, so that also $\delta_j \downarrow 0$ and $\alpha_j \downarrow p(u, v)$. Let

$$A_j = \{x' \mid \operatorname{dist}(x', A) \le \epsilon_j\},$$
$$B_j = \{(u', v') \mid p(u', v') < \alpha_j \quad \text{and} \quad |(u', v') - (u, v)| < \delta_j\}.$$

Then A_j is a compact set such that for every $(u', v') \in B_j$ (and in particular for $(u', v') = (u, v)$), problem $(\mathrm{P}_{u',v'})$ has feasible solutions in A_j, and over these the infimum of $f_0(v', \cdot)$ is still $p(u', v')$. Since this restricted infimum concerns a continuous function over a certain set that is nonempty and compact (because A_j is compact, D is closed, and every f_i is continuous), it is attained at some point.

Thus when $(u', v') \in B_j$, there is an optimal solution to $(P_{u',v'})$ in A_j (and for $(u', v') = (u, v)$ there is an optimal solution to $(P_{u,v})$ in $\bigcap_j A_j = A$). For each j and each $(u', v') \in B_j$ with $(u', v') \notin B_{j+1}$, select some optimal solution on $(P_{u',v'})$ in A_j and denote it by $\xi(u', v')$; let $\xi(u, v)$ itself denote some optimal solution to $(P_{u,v})$ in A. Then ξ is a well-defined mapping on the set (5.5) (identical to B_0 in the present notation), and $\xi(u', v') \in A_j$ when $(u', v') \in B_j$. This mapping meets all prescriptions: inasmuch as $\epsilon_j \downarrow 0$, $\delta_j \downarrow 0$, and $\alpha_j \downarrow p(u, v)$, all cluster points of $\xi(u', v')$ as $(u', v') \to_p (u, v)$ are contained in $\bigcap_j A_j = A$.

Remark. The tameness condition we have been exploring was inspired in part by a condition introduced by Hiriart-Urruty [17] in a related context. This is clarified by the equivalence in Proposition 10. Hiriart-Urruty's condition is essentially the one in Proposition 10, but stronger in having ordinary topology appear in place of the '$\to_p$' topology.

Other authors who have dealt with this subject have relied on still more stringent assumptions. For instance, to follow the pattern of the papers of Gauvin and Tolle [13], Gauvin [11], Gauvin and Dubeau [12], the multifunction that associates to each (u', v') the set of all feasible solutions to $(P_{u',v'})$ would be assumed to be bounded on an ordinary neighborhood of (u, v). See also earlier work of Evans and Gould [9], Greenberg and Pierskalla [16], on upper and lower semicontinuity properties of optimal value functions.

In our main theorem, which we are now ready to present, 'co' denotes convex hull and 'cl' closure. Again we use the concept of 'pointedness', as defined in Section 2 for cones that are not necessarily convex.

Theorem 2. *Suppose* (u, v) *is such that* $(P_{u,v})$ *is tame, and let* X *be any set of optimal solutions to* $(P_{u,v})$ *that at least includes whatever optimal solutions to* $(P_{u,v})$ *happen to lie in* A*, the set invoked in the definition* (5.1) *of tameness. (In particular,* X *could be taken to be the set of all optimal solutions to* $(P_{u,v})$*.) Then*

$$\partial p(u, v) = \operatorname{cl}\operatorname{co}\Big\{\Big(\Big[\bigcup_{x\in X} K(u, v, x)\Big] \cap \partial p(u, v)\Big) + \Big[\bigcup_{x\in X} K_0(u, v, x)\Big] \cap \partial^0 p(u, v)\Big\}, \tag{5.6}$$

$$\partial^0 p(u, v) \supset \operatorname{cl}\operatorname{co}\Big\{\Big[\bigcup_{x\in X} K_0(u, v, x)\Big] \cap \partial^0 p(u, v)\Big\}. \tag{5.7}$$

Equality holds in (5.7) *if* $\bigcup_{x\in X} K(u, v, x) \cap \partial p(u, v) = \emptyset$*, or if the cone*

$$\Big[\bigcup_{x\in X} K(u, v, x)\Big] \cap \partial^0 p(u, v)$$

is pointed; in the latter case $\partial^0 p(u, v)$ *too is pointed, and the closure operation is superfluous in both* (5.6) *and* (5.7).

Although the proof of Theorem 2 will not be laid out until Section 6, we shall proceed immediately with some corollaries. Consequences about directional derivatives will be saved for Section 7. The reader should note, incidentally, that Theorem 2 and everything that will be based on it remain valid if $K(u, v, x)$ and $K_0(u, v, x)$ are replaced by other sets that at least are sure to include them. In particular, the multiplier conditions (4.4) and (4.5) could be supplanted by the slightly weaker ones of Clarke [3] or Hiriart-Urruty [18] corresponding to the inclusions in Proposition 6.

Corollary 1. *Assuming tameness as in Theorem 2, one has*

$$\partial p(u, v) \subset \operatorname{cl}\operatorname{co}\Big\{ \bigcup_{x\in X} K(u, v, x) + \bigcup_{x\in X} K_0(u, v, x)\Big\}. \tag{5.8}$$

If in addition the cone $\bigcup_{x\in X} K_0(u, v, x)$ *is pointed, then* $\partial^0 p(u, v)$ *is pointed and*

$$\partial^0 p(u, v) \subset \operatorname{cl}\operatorname{co}\Big\{ \bigcup_{x\in X} K_0(u, v, x)\Big\}. \tag{5.9}$$

Corollary 2. *Assuming tameness as in Theorem 2, suppose* $K_0(u, v, x) \cap \partial^0 p(u, v) = \{(0, 0)\}$ *for all* $x \in X$ *(as is true certainly if every optimal solution* x *to* $(P_{u,v})$ *satisfies the constraint qualification* $K_0(u, v, x) = \{(0, 0)\}$*). Then* p *is locally Lipschitzian on a neighborhood of* (u, v) *and*

$$\partial p(u, v) = \operatorname{cl}\operatorname{co}\Big\{ \Big[\bigcup_{x\in X} K(u, v, x)\Big] \cap \partial p(u, v)\Big\}; \tag{5.10}$$

in particular,

$$\partial p(u, v) \subset \operatorname{cl}\operatorname{co}\Big\{ \bigcup_{x\in X} K(u, v, x)\Big\}, \tag{5.11}$$

$$\partial_v p(u, v) \subset \operatorname{cl}\operatorname{co}\Big\{ z \;\Big|\; \exists y \text{ with } (y, z) \in \bigcup_{x\in X} K(u, v, x)\Big\}. \tag{5.12}$$

This follows via Propositions 3 and 4. It encompasses the estimate of Gauvin [11] for the smooth case (a), namely: under the assumption that $p(u, v) < \infty$ and

$$\{x \mid (u', v') \text{ with } x \text{ feasible in } (P_{u',v'}),\ |(u', v') - (u, v)| \le \delta\}$$
$$\text{is a compact set for some } \delta > 0, \tag{5.13}$$

if every optimal solution x to $(P_{u,v})$ satisfies the Mangasarian–Fromovitz constraint qualification (4.16), then p is locally Lipschitzian on a neighborhood of (u, v) and (5.11) holds. Corollary 2 also covers the estimate of Gauvin and Dubeau [12], which is (5.12) under the same assumptions. Of course (5.10) is a stronger assertion than (5.12), and Corollary 2 shows that it is valid under far more general circumstances than established previously. Corollary 1, on the other hand, shows that Theorem 2 yields outer estimates for $\partial p(u, v)$ even in cases where p is not locally Lipschitzian around (u, v). This is a new level of result.

Outer estimates for $\partial_v p(u, v)$ more subtle than (5.12) can be produced by combining Proposition 4 directly with Theorem 2. We leave the details to the reader.

Corollary 3. *Assuming tameness as in Theorem 2, one has for the closed convex cone*

$$G = \bigcup_{x\in X} \{(h, k) \mid y \cdot h + z \cdot k \le 0 \text{ for all } (y, z) \in K_0(u, v, x)\} \tag{5.14}$$

that p is directionally Lipschitzian with respect to every $(h, k) \in \operatorname{int} G$.

Proof. Apply Proposition 2 and Corollary 1.

Corollary 4. *Assuming tameness as in Theorem 2, if* $\partial p(u, v) \neq \emptyset$ *(as is true in particular whenever p is calm at* (u, v), *cf. Proposition* 1), *then* $(P_{u,v})$ *has an optimal solution* $x \in X$ *for which there is a vector* $(y, z) \in K(u, v, x)$ *that belongs to* $\partial p(u, v)$.

This result demonstrates that Theorem 2 yields not only 'outer estimates' but 'inner estimates'. Corollary 4 extends a result of Clarke and Aubin [6] for problem $(P_{u,v})$ in the 'almost convex' case, where everything is as in case (b) of Section 1 except that the objective function f_0 is not necessarily convex. It also covers a somewhat more general result of Aubin [1], although the connection in this case takes more effort to establish. The results in question are posed in terms of a problem structure that is different from the one in $(P_{u,v})$, although ultimately encompassed by it. However they also apply to infinite-dimensional problems, in contrast to Corollary 4.

The results in our earlier paper [29] can also be mentioned in conjunction with Corollary 4. These show the existence in $\partial p(u, v)$ of certain limits of multiplier vectors that satisfy higher-order optimality conditions.

Corollary 5. *Under the hypothesis of Theorem 2, if X is a singleton* $\{x\}$, *and for this x the set* $K(u, v, x)$ *is a singleton* $\{(y, z)\}$ *and the constraint qualification* $K_0(u, v, x) = \{(0, 0)\}$ *is satisfied, then p is strictly differentiable at* (u, v) *with* $\nabla p(u, v) = (y, z)$.

Proof. The assumptions imply via Theorem 2 that $\partial p(u, v) = \{(y, z)\}$, and this is equivalent to p being strictly differentiable at (u, v) with $\nabla p(u, v) = (y, z)$, as already noted in Section 2.

Remark. The constraint qualification in Corollary 5 does not have to be postulated separately in cases (a), (b), (c), (d) or (e) of $(P_{u,v})$. In those cases it follows from $K(u, v, x)$ being a singleton; cf. Proposition 7.

Corollary 6. *Suppose there is a mapping* ξ *as described in Proposition* 10, *and let X be the set of all cluster points of* $\xi(u', v')$ *as* $(u', v') \to_p (u, v)$ *(in the sense of* (2.2)). *Then the hypothesis of Theorem* 2 *is satisfied, so the conclusions in Theorem* 2 *and Corollaries* 1, 3 *and* 4 *(and under extra assumptions about X the conclusions in Corollaries* 2 *and* 5) *are valid.*

Proof. This follows from Proposition 10.

Corollary 7. *Suppose either that* $(P_{u,v})$ *is tame and has a unique optimal solution* x, *or that* $(P_{u,v})$ *has an optimal solution* x *that can be perturbed continuously in the sense of the existence of a mapping* ξ *as in Proposition* 10 *with* $\xi(u',v')\to x$ *as* $(u',v')\to_p(u,v)$. *If* $(P_{u,v})$ *falls into the subdifferentially regular case* (d) *or extended smooth case* (e) *in Section* 4 (*or in particular one of cases* (a), (b) *and* (c) *of Section* 1), *then*

$$\partial p(u,v)\subset K(u,v,x) \quad and \quad \partial^0 p(u,v)\subset K_0(u,v,x). \tag{5.15}$$

Proof. Either way, we can apply Theorem 2 with $X=\{x\}$ (cf. Corollary 6). Furthermore, the conclusions of Proposition 7 hold in their strongest form. The formulas in Theorem 2 then reduce to (5.15) by virtue of $\partial p(u,v)$ and $\partial^0 p(u,v)$ being closed and convex, with $\partial^0 p(u,v)$ equal to the recession cone of $\partial p(u,v)$ unless $\partial p(u,v)=\emptyset$.

6. The main arguments

We proceed now to derive Theorem 1 from Theorem 2 using a certain characterization of our calmness property, and then to prove Theorem 2 itself by means of a new general result about limits of subgradients.

Proposition 11. *Let* x *be a locally optimal solution to* $(P_{u,v})$. *Let* $\theta:[0,\infty)\to[0,\infty)$ *be any increasing convex function with* $\theta(0)=0$ *and* $\theta'(0)=0$, *and let* $\epsilon>0$. *Then the parameterized problem*

$$(\tilde{P}_{u',v'})\qquad \begin{array}{l} \text{minimize } \tilde{f}_0(v',x')=f_0(v',x')+\theta(|x'-x|) \text{ over all } x' \text{ satisfying} \\ (v',x')\in\tilde{D}=\{(v',x')\in D \mid |x'-x|\le\epsilon\} \text{ and} \\ f_i(v',x')+u_i'\begin{cases}\le 0 & \text{for } i=1,\ldots,s,\\ =0 & \text{for } i=s+1,\ldots,m\end{cases}\end{array}$$

in place of $(P_{u',v'})$ *again satisfies the fundamental assumptions of Section* 1: $\tilde{D}$ *is again closed and* $\tilde{f}_0$ *locally Lipschitzian. Moreover, the term* $g(x')=\theta(|x'-x|)$ *in* $\tilde{f}_0$ *is a finite convex function of* x' (*therefore locally Lipschitzian*) *which is strictly differentiable at* $x'=x$ *with*

$$\nabla g(x)=0, \qquad g(x)=0, \qquad g(x')>0 \quad \text{for} \quad x'\neq x. \tag{6.1}$$

Furthermore, if $\epsilon<\rho$, *where* ρ *is the radius of a spherical neighborhood of* x *with respect to which the local optimality of* x *holds in* $(P_{u,v})$, *then* x *is the unique (globally) optimal solution to* $(\tilde{P}_{u,v})$, *and the optimal value function*

$$\tilde{p}(u',v')=\inf(\tilde{P}_{u',v'}) \tag{6.2}$$

has $\tilde{p}(u,v)=f_0(v,x)$. *If in addition* x *is actually a globally optimal solution to*

$(\mathrm{P}_{u,v})$, *then*

$$\tilde{p}(u,v)=p(u,v), \quad \textit{while } \tilde{p}(u',v')\geq p(u',v') \quad \textit{for all } (u',v')\neq(u,v). \tag{6.3}$$

Proof. All these assertions are elementary, except for the differentiability property of g. The convexity of g allows us to compute $g'(x;h)=\theta'(0)|h|=0$ for all h, from which it follows (cf. [26, Section 23]) that $\partial g(x)=\{0\}$. Then g must be strictly differentiable at 0 with $\nabla g(x)=0$, according to the results cited in Section 2.

Proposition 12. *Let x be a locally optimal solution to $(\mathrm{P}_{u,v})$. For $(\mathrm{P}_{u,v})$ to be calm at x, it is necessary and sufficient that for every function θ as in Proposition* 11, *one has for all $\epsilon>0$ sufficiently small that the modified optimal value function $\tilde{p}$ in Proposition* 11 *is calm at (u,v) (in the sense of Proposition* 1).

Proof. The argument will utilize the notation and conclusions of Proposition 11.

Necessity. Suppose $(\mathrm{P}_{u,v})$ is calm at x, and fix any θ as described. If for some $\epsilon\in(0,\rho)$ the function $\tilde{p}$ is not calm at (u,v), there exist for any $\beta\in\mathbb{R}$ points (u',v') arbitrarily near to (u,v) and yielding

$$[\tilde{p}(u',v')-\tilde{p}(u,v)]/|(u',v')-(u,v)|<\beta.$$

Here $\tilde{p}(u,v)=f_0(v,x)$, so the inequality means by the definition of $\tilde{p}$ that

$$[f_0(v',x')+g(x')-f_0(v,x)]/|(u',v')-(u,v)|<\beta$$

for some feasible solution x' to $(\mathrm{P}_{u',v'})$ with $|x'-x|\leq\epsilon$. Thus if there is a sequence of values $\epsilon_j\downarrow 0$ such that the corresponding functions $\tilde{p}$ are not calm at (u,v), we can select for any sequence of values $\beta_j\downarrow-\infty$, corresponding points (u^j,v^j) arbitrarily near to (u,v) and feasible solutions x^j to (P_{u^j,v^j}) with $|x^j-x|\leq\epsilon_j$ and

$$[f_0(v^j,x^j)+g(x^j)-f_0(v,x)]/|(u^j,v^j)-(u,v)|<\beta_j.$$

Then $x^j\to x$ and (since $g\geq 0$)

$$[f_0(v^j,x^j)-f_0(v,x)]/|(u^j,v^j)-(u,v)|\to-\infty. \tag{6.4}$$

In particular, (u^j,v^j) can be selected so as to converge to (u,v), and a contradiction is then obtained to the assumption that $(\mathrm{P}_{u,v})$ is calm at x. Hence under this assumption there cannot exist a sequence of values $\epsilon_j\downarrow 0$ such that $\tilde{p}$ is not calm at (u,v), and this is what we needed to prove.

Sufficiency. Suppose $(\mathrm{P}_{u,v})$ is not calm at x. Then there exist $(u^j,v^j)\to(u,v)$ and $x^j\to x$ such that x^j is feasible in (P_{u^j,v^j}) and (6.4) holds. Let $\delta_j=|(u^j,v^j)-(u,v)|$ and $\epsilon_j=|x^j-x|$; passing to subsequences if necessary, we can suppose that δ_j and ϵ_j are strictly decreasing in j. The line segment joining the points $(\epsilon_j, \delta_j\epsilon_j)$ and $(\epsilon_{j+1}, \delta_{j+1}\epsilon_{j+1})$ in $\mathbb{R}^2$ has slope

$$\lambda_j = (\delta_j\epsilon_j - \delta_{j+1}\epsilon_{j+1})/(\epsilon_j - \epsilon_{j+1})$$

which satisfies

$$\delta_j > \lambda_j > \delta_{j+1}. \tag{6.5}$$

Let $\theta : [0, \infty) \to [0, \infty)$ be the function whose graph is the union of all these segments and $(0, 0)$; then θ is continuous with

$$\theta(\epsilon_j) = \delta_j\epsilon_j \quad \text{for all } j, \qquad \theta(0) = 0.$$

From (6.5) we have $\lambda_j > \lambda_{j+1} > \cdots > 0$; hence θ is actually convex and increasing. Thus θ belongs to the class of functions under consideration, and

$$\theta(|x^j - x|)/|(u^j, v^j) - (u, v)| = \epsilon_j \to 0. \tag{6.6}$$

It will be demonstrated now that for this θ and any $\epsilon \in (0, \rho)$, the function $\tilde{p}$ is not calm at (u, v). Indeed, since x^j is a feasible solution to (P_{u^j, v^j}) with $|x^j - x| = \epsilon_j$, it is a feasible solution to $(\tilde{\mathrm{P}}_{u^j, v^j})$ for j sufficiently large, and then

$$\tilde{p}(u^j, v^j) \le f_0(v^j, x^j) + \theta(|x^j - x|).$$

We also have $\tilde{p}(u, v) = f_0(v, x)$ (because $\epsilon < \rho$), and since (6.4) holds for the chosen sequences, we see via (6.6) that

$$\frac{\tilde{p}(u^j, v^j) - \tilde{p}(u, v)}{|(u^j, v^j) - (u, v)|} \le \frac{f_0(v^j, x^j) - f_0(v, x)}{|(u^j, v^j) - (u, v)|} + \epsilon_j \to -\infty.$$

Thus

$$\liminf_{(u', v') \to (u, v)} \frac{\tilde{p}(u', v') - \tilde{p}(u, v)}{|(u', v') - (u, v)|} = -\infty$$

as claimed.

Proof of Theorem 1 (using Theorem 2). Suppose x is a locally optimal solution to $(\mathrm{P}_{u,v})$. Then for θ and ϵ as in Proposition 11 with $\epsilon < \rho$, x is the unique optimal solution to the modified problem $(\tilde{\mathrm{P}}_{u,v})$. Theorem 2 can be applied to $(\tilde{\mathrm{P}}_{u,v})$ (cf. Proposition 9). Taking $X = \{x\}$, we conclude in particular that

$$\partial\tilde{p}(u, v) \subset \mathrm{cl}\,\mathrm{co}\{\tilde{K}(u, v, x) + \tilde{K}_0(u, v, x)\}, \quad \text{and if } \tilde{K}_0(u, v, x) \text{ is pointed, also } \partial^0\tilde{p}(u, v) \subset \mathrm{cl}\,\mathrm{co}\,\tilde{K}_0(u, v, x), \tag{6.7}$$

where $\tilde{K}(u, v, x)$ and $\tilde{K}_0(u, v, x)$ are the multiplier sets corresponding to $(\tilde{\mathrm{P}}_{u,v})$, i.e., with $\tilde{f}_0$ and $\tilde{D}$ is place of f_0 and D. Actually

$$\tilde{K}(u, v, x) = K(u, v, x) \quad \text{and} \quad \tilde{K}_0(u, v, x) = K_0(u, v, x). \tag{6.8}$$

because

$$\partial\left[\tilde{f}_0 + \sum_{i=1}^{m} y_i f_i + \delta_{\tilde{D}}\right](v, x) = \partial\left[f_0 + \sum_{i=1}^{m} y_i f_i + \delta_D\right](v, x);$$

this is due to the fact that $\tilde{D}$ and D coincide in a neighborhood of (v, x), while $\tilde{f}_0$ and f_0 differ only by a function whose gradient vanishes at (v, x); cf. Proposition 5.

If $K_0(u, v, x) = \{(0, 0)\}$, we have $0^+K(u, v, x) = \{(0, 0)\}$ (Proposition 7), so $K(u, v, x)$ is not just closed (Proposition 7) but compact. Then too, $\partial^0 p(u, v) = \{(0, 0)\}$ by (6.7) and (6.8). Hence according to Proposition 3, $\tilde{p}$ is Lipschitzian around (u, v) and in particular calm at (u, v). Since this holds regardless of the choice of θ, as long as $\epsilon \in (0, \rho)$, it is clear from Proposition 12 that $(P_{u,v})$ must be calm at x.

However, if $(P_{u,v})$ is calm at x, we know from Proposition 12 that when ϵ is sufficiently small, $\tilde{p}$ is calm at (u, v) and therefore by Proposition 1 that $\partial p(u, v) \neq \emptyset$. Then we deduce from (6.7) and (6.8) that $K(u, v, x) \neq \emptyset$. This completes the derivation of Theorem 1.

Next on the agenda is the proof of Theorem 2. The following result is the first step.

Proposition 13. *Suppose (u, v) is such that $(P_{u,v})$ is tame, and let A be a set for which the definition (5.1) of tameness is fulfilled. Then there are numbers $\delta > 0$ and $\alpha > p(u, v)$ and a compact set $\hat{D} \subset D$, such that the replacement of D by $\hat{D}$ does not affect the infimum $p(u', v')$ for any (u', v') satisfying $|(u', v') - (u, v)| < \delta$ and $p(u', v') < \alpha$, nor does it alter the set of optimal solutions x to $(P_{u,v})$ which lie in A or the sets $K(u, v, x)$ and $K_0(u, v, x)$ associated with any such x.*

Proof. Fix any $\epsilon > 0$ and corresponding δ and α as in (5.1). Let

$$\hat{D} = \{(v', x') \in D \mid \operatorname{dist}(x', A) \leq \epsilon \text{ and } |v' - v| \leq \delta\}.$$

Since A is compact and D is closed, $\hat{D}$ is compact. The assertions are then all obvious from (5.1) and the fact that the sets $\{x \mid (v, x) \in \hat{D}\}$ and $\{x \mid (v, x) \in D\}$ agree in a neighborhood of A.

We will also need a new general result about convergence of subgradients.

Proposition 14. *Suppose for $j = 1, 2, \ldots$, that x^j furnishes a finite local minimum of $f + g_j$, where f and g_j are lower semicontinuous functions from $\mathbb{R}^n$ to $(-\infty, \infty]$. If $x^j \to x$, $f(x^j) \to f(x)$ (finite) and $\partial g_j(x_j) \to \{0\}$ (in the sense that for every neighborhood U of 0 one has $\emptyset \neq \partial g_j(x^j) \subset U$ when j is sufficiently large), then $0 \in \partial f(x)$.*

Proof. Suppose $0 \notin \partial f(x)$. Then there exists a vector h such that $f^\uparrow(x; h) < 0$, i.e.,

$$0 > \lim_{\epsilon \downarrow 0} \limsup_{\substack{x' \to_f x \\ t \downarrow 0}} \left[\inf_{|h' - h| \leq \epsilon} \frac{f(x' + th') - f(x')}{t} \right]$$

(cf. the general formula (2.5) for subgradients in terms of subderivatives). In particular, for every $\epsilon > 0$ and sequence $t_j \downarrow 0$ one has

$$0 > \limsup_{j\to\infty} \left[\inf_{|h'-h|\le\epsilon} \frac{f(x^j + t_j h') - f(x^j)}{t_j} \right]. \tag{6.9}$$

The assumption that $f + g_j$ has a local minimum at x^j implies

$$f(x^j + t_j h') + g_j(x^j + t_j h') \ge f(x^j) + g_j(x^j)$$

for any h' once j is large enough, and this can be written

$$[f(x^j + t_j h') - f(x^j)]/t_j \ge -[g_j(x^j + t_j h') - g_j(x^j)]/t_j.$$

Hence by (6.9), for every $\epsilon > 0$ and sequence $t_j \downarrow 0$, one has

$$0 < \liminf_{j\to\infty} \left[\sup_{|h'-h|\le\epsilon} \frac{g_j(x^j + t_j h') - g_j(x^j)}{t_j} \right]. \tag{6.10}$$

Next we use the fact that $\partial g_j(x^j) \to \{0\}$; passing to a subsequence if necessary, it can be supposed that

$$\emptyset \ne \partial g_j(x^j) \subset \{w \mid |w| < \lambda_j\} \quad \text{where } \lambda_j \downarrow 0. \tag{6.11}$$

In particular $\partial g_j(x^j)$ is bounded; hence g_j is locally Lipschitzian around x^j [25, Theorem 4]. Moreover, (6.11) implies that for all $k \in \mathbb{R}^n$

$$\begin{aligned} \lambda_j|k| &> \max\{w \cdot k \mid w \in \partial g_j(x^j)\} \\ &= g_j^\circ(x^j; k) = \limsup_{\substack{x'\to x^j \\ t\downarrow 0}} \frac{g_j(x' - tk) - g_j(x')}{t} \end{aligned}$$

(recall from Section 2 that $g_j^\uparrow(x^j; k)$ reduces to $g_j^0(x^j; k)$ in the locally Lipschitzian case). Thus λ_j serves as a Lipschitz constant for g_j in some neighborhood of x_j, say in a ball of radius δ_j around x^j. Fixing $\epsilon > 0$ arbitrarily, choose the sequence $t_j \downarrow 0$ so that $x^j + t_j h'$ belongs to this neighborhood for all h' satisfying $|h' - h| \le \epsilon$ (it suffices to have $0 < t_j < \delta_j/\epsilon$). Then for all j sufficiently large one has

$$g_j(x^j + t_j h') - g_j(x^j) \le t_j \lambda_j |h'| \quad \text{whenever } |h' - h| \le \epsilon,$$

and hence from (6.10)

$$0 < \liminf_{j\to\infty} \lambda_j(|h| + \epsilon).$$

This contradicts the fact that $\lambda_j \downarrow 0$ and establishes that $0 \in \partial f(x)$ after all.

Proof of Theorem 2. Proposition 13 gives us license to suppose without loss of generality that D is a compact set. It is elementary then that $(\mathrm{P}_{u',v'})$ has an optimal solution for every (u', v') such that it has a feasible solution (in particular for (u, v), because $p(u, v) < \infty$ by hypothesis), and that p is lower semicontinuous everywhere and globally bounded below. By means of the reformulation in Section 1, we can also reduce everything to the notationally simpler case where there are no vectors v and z. The reasoning here, as far as the equivalence of the multiplier conditions is concerned, is based on Proposition 5: the reformulation involves the introduction merely of linear constraint functions, and $\partial(g_1+g_2)=\nabla g_1+\partial g_2$ in particular when g_1 is linear.

In this reduced case with D compact, a formula proved in [29, Theorem 2] in terms of the (quadratic) augmented Lagrangian function becomes applicable:

$$\partial p(u)=\operatorname{cl}\operatorname{co}[Y+Y_0] \quad \text{and} \quad \partial^0 p(u)=\operatorname{cl}\operatorname{co} Y_0, \tag{6.12}$$

where

$$Y=\{y \mid \exists u^j \to_p u \text{ and } y^j \text{ an augmented multiplier vector for } (\mathrm{P}_{u^j}), \text{ such that } y^j \to y\}, \tag{6.13}$$

$$Y_0=\{y \mid \exists\, u^j \to_p u,\ \lambda_j \downarrow 0 \text{ and } y^j \text{ an augmented multiplier vector for } (\mathrm{P}_{u^j}), \text{ such that } \lambda_j y^j \to y\}. \tag{6.14}$$

Here y^j is called an 'augmented multiplier vector' for (P_{u^j}) if for all $r_j>0$ sufficiently large, the optimal solutions to (P_{u^j}) are precisely the vectors x^j such that (x^j, y^j) is a (global) saddlepoint of the augmented Lagrangian

$$L(u^j, x, y, r_j)=f_0(x)+\frac{1}{2r_j}\sum_{i=1}^{s}[y_i+r_j(f_i(x)+u_i^j)]_+^2 + \frac{1}{2r_j}\sum_{i=s+1}^{m}[y_i+r_j(f_i(x)+u_i^j)]^2-\frac{1}{2r_j}|y|^2$$

with respect to $x\in D$ and $y\in\mathbb{R}^m$. (We are using the notation that $s_+=\max\{s,0\}$.) Since $\partial p(u)$ and $\partial^0 p(u)$ are closed convex sets, we will be able to derive formulas (5.6) and (5.7) in Theorem 2 by showing that

$$Y\subset\bigcup_{x\in X}K(u,x) \quad \text{and} \quad Y_0\subset\bigcup_{x\in X}K_0(u,x). \tag{6.15}$$

Consider now any sequences $y^j\to y$ and $u^j\to_p u$ such as in the definition (6.13) of Y. Since D is compact and definition (5.1) is fulfilled by A, there is (for j sufficiently large) a sequence of points x^j such that x^j is an optimal solution to (P_{u^j}) and $\operatorname{dist}(x^j, A)\to 0$. Passing to a subsequence if necessary, we can suppose x^j converges to some $x\in A$. Then x is an optimal solution to (P_u) (by the continuity of the functions f_i and the fact that $p(u^j)\to p(u)$), and x belongs to X (the hypothesized set of optimal solutions to (P_u) which includes all those in A). We will demonstrate that $y\in K(u,x)$, and this will establish the first inclusion in (6.15).

To say that (x^j, y^j) is a saddlepoint of $L(u^j, x, y, r_j)$ with respect to $x\in D$ and $y^j\in\mathbb{R}^m$ is to say that

$$f_i(x^j)+u_i^j\le 0, \qquad y_i^j\ge 0, \qquad y_i^j[f_i(x^j)+u_i^j]=0 \quad \text{for } i=1,\dots,s,$$
$$f_i(x^j)+u_i^j=0 \quad \text{for } i=s+1,\dots,m, \tag{6.16}$$

and that x^j gives the global minimum over D of $L(u^j, y^j, r_j)$. But (6.16) implies that the latter function reduces locally around x^j to

$$f_0+\sum_{i=1}^{m} y_i^j[f_i+u_i^j]+\frac{r_j}{2}\Big(\sum_{i\in I_0(j)}[f_i+u_i^j]_+^2+\sum_{i\in I_1(j)}[f_i+u_i^j]^2\Big), \tag{6.17}$$

where

$$\begin{aligned} I_0(j) &= \text{set of all } i\in\{1,\dots,s\} \text{ with } y_i^j=0,\\ I_1(j) &= \text{set of all other constraint indices.} \end{aligned} \tag{6.18}$$

This reduction makes use of the relation

$$\frac{1}{2r_j}([y_i^j+r_j(f_i+u_i^j)]_+^2-(y_i^j)^2)$$

$$=\begin{cases} y_i^j[f_i+u_i^j]+\frac{r_j}{2}[f_i+u_i^j]^2 & \text{where } f_i+u_i^j\geq -y_i^j/r_j,\\ -\frac{1}{2r_j}(y_i^j)^2 & \text{where } f_i+u_i^j\leq -y_i^j/r_j. \end{cases} \tag{6.19}$$

(For active inequality constraint indices i with $y_i^j>0$, one has $-y_i^j/r_j<0$ but $f_i(x^j)+u_i^j=0$, so only the first alternative in (6.19) holds in a certain neighborhood of x^j. For all other inequality constraint indices one has $y_i^j=0$, so that (6.19) simply gives $[f_i+u_i^j]_+^2$.)

From (6.16) we know that in the limit as $j\to\infty$:

$$\begin{aligned} &f_i(x)+u_i\leq 0, \quad y_i\geq 0, \quad y_i[f_i(x)+u_i]=0 \quad \text{for } i=1,\dots,s,\\ &f_i(x)+u_i=0 \quad \text{for } i=s+1,\dots,m. \end{aligned} \tag{6.20}$$

On the other hand, we have seen that x^j gives a local minimum to the function (6.18) over D. This tells us that x^j gives a local minimum to $f+g_j$, where

$$f=f_0+\sum_{i=1}^{m} y_if_i+\delta_D, \tag{6.21}$$

$$g_j=\sum_{i=1}^{m}(y_i^j-y_i)f_i+\frac{r_j}{2}\sum_{i=1}^{m}h_{ij}^2, \tag{6.22}$$

$$h_{ij}=\begin{cases}[f_i+u_i^j]_+ & \text{for } i\in I_0(j),\\ f_i+u_i^j & \text{for } i\in I_1(j).\end{cases} \tag{6.23}$$

The functions h_{ij} all vanish at x^j by virtue of the definitions (6.18), and these functions are all locally Lipschitzian. In applying to (6.22) the rules of subdifferential calculus for sums and squares (cf. Proposition 5 and [5, Section 13]), we get

$$\partial g_j(x^j)\subset\sum_{i=1}^{m}(y_i^j-y_i)\partial f_i(x^j)+r_j\sum_{i=1}^{m}h_{ij}(x^j)\partial h_{ij}(x^j),$$

where $h_{ij}(x^j)=0$; the second sum therefore drops out. But $\lim_j(y_i^j-y_i)=0$ and $\limsup_j\partial f_i(x^j)\subset\partial f_i(x)$ (because f_i is locally Lipschitzian and $x^j\to x$, cf. [3]).

Hence $\partial g_j(x^j) \to \{0\}$, and we may conclude from Proposition 14 that $0 \in \partial f(x)$ for f as in (6.21). This property along with (6.16) means that $y \in K(u, x)$ as claimed.

The argument is very similar in the case of y^j, u^j and λ_j such as in the definition (6.14) of Y_0. The difference comes in multiplying (6.17) through by λ_j and characterizing x^j accordingly as a local minimizer of $f + g_j$ taken as

$$f = \sum_{i=1}^{m} y_i f_i + \delta_D,$$

$$g_j = \lambda_j f_0 + \sum_{i=1}^{m} (\lambda_j y_i^j - y_i) f_i + \frac{r_j}{2} \sum_{i=1}^{m} h_{ij}^2,$$

with h_{ij} as in (6.3). Again $\partial g_j(x^j) \to \{0\}$, so $0 \in \partial f(x)$ by Proposition 14, and the conclusion is obtained that $y \in K_0(u, x)$.

Thus the second inclusion in (6.15) is valid too, and formulas (5.6) and (5.7) of Theorem 2 are then true in consequence of (6.12), as already explained.

To obtain via (6.12) and (6.15) the final assertions of Theorem 2, about equality holding in (5.7), it will suffice to prove that

$$\partial^0 p(u) = \text{cl co } Y_0 \tag{6.24}$$

if either $Y = \emptyset$ or Y_0 is pointed, and that in the latter case one actually has

$$\partial p(u) = \text{co}[Y + Y_0] \quad \text{and} \quad \partial^0 p(u) = \text{co } Y_0. \tag{6.25}$$

Here we must delve deeper into the argument in [29] by means of which (6.12) was established. The argument was based on representing $\partial p(u)$ and $\partial^0 p(u)$ in terms of the cone

$$N = \{\lambda(y, -1) \mid y \in Y, \lambda > 0\} \cup \{(y, 0) \mid y \in Y_0\} \tag{6.26}$$

in $R^m \times R$ by the formulas

$$\partial p(u) = \{y \mid (y, -1) \in \text{cl co } N\}, \quad \partial^0 p(u) = \{y \mid (y, 0) \in \text{cl co } N\}. \tag{6.27}$$

(See [29, Theorem and its proof].) It was observed also that

$$0 \in Y_0 \supset 0^+ Y := \{y \mid \exists y^j \in Y, \lambda_j \downarrow 0, \text{ with } \lambda_j y^j \to y\}, \tag{6.28}$$

or what amounts to the same thing, that N is closed and nonempty. The statements about (6.24) and (6.25) at the beginning of this paragraph, as well as (6.12) itself, are implied by this representation, as we demonstrate in the following geometric proposition, thereby completing proof of Theorem 2.

Proposition 15. *Let Y and Y_0 be any closed subsets of R^m such that Y_0 is a cone satisfying (6.28), and let N be the closed cone in $R^m \times R$ defined by (6.26). Then*

$$\{y \mid (y, -1) \in \text{cl co } N\} = \text{cl co}[Y + Y_0], \tag{6.29}$$

$$\{y \mid (y, 0) \in \text{cl co } N\} \supset \text{cl co } Y_0. \tag{6.30}$$

Equality holds in (6.30) *if* $Y = \emptyset$, *or if* Y_0 *is pointed; in the latter case* co Y_0 *is itself closed and pointed, as is* co N, *and one actually has*

$$\{y \mid (y, -1) \in \text{cl co } N\} = \text{co}[Y + Y_0], \tag{6.31}$$

$$\{y \mid (y, 0) \in \text{cl co } N\} = \text{co } Y_0. \tag{6.32}$$

Proof. It is trivial from (6.26) that (6.30) always holds, and that it holds with equality when $Y = \emptyset$. Note too that when $Y = \emptyset$, (6.29) and (6.31) hold with both sides empty. We can therefore suppose henceforth that $Y \neq \emptyset$. Then N meets both of the open half-spaces bounded by the hyperplane $H = \{(y, -1) \mid y \in R^m\}$, so co N certainly cannot be separated from H and hence $H \cap \text{ri co } N \neq \emptyset$ [3, Section 11]. This implies

$$H \cap \text{cl co } N = \text{cl}[H \cap \text{co } N]$$

[3, Section 6], or equivalently,

$$\{y \mid (y, -1) \in \text{cl co } N\} = \text{cl}\{y \mid (y, -1) \in \text{co } N\}. \tag{6.33}$$

On the other hand, since Y_0 is a cone containing 0, we find from (6.26) that

$$\{y \mid (y, -1) \in \text{co } N\} = \text{co}[Y + Y_0], \tag{6.34}$$

and of course

$$\{y \mid (y, 0) \in \text{co } N\} = \text{co } Y_0. \tag{6.35}$$

The combination of (6.33) and (6.34) yields (6.29).

We shall demonstrate now that if Y_0 is pointed, then co Y_0 is closed and pointed. Since N, like Y_0, is a closed cone containing the origin, and since N obviously is pointed if and only if Y_0 is pointed, co N too is closed and pointed. Then (6.31) and (6.32) will be seen simply as restatements of (6.34) and (6.35).

Assume Y_0 is pointed. Because Y_0 is a cone in R^m containing the origin, we have (by Carathéodory's theorem [3, Section 17])

$$\text{co } Y_0 = \{y^1 + \cdots + y^{m+1} \mid y^k \in Y_0\}.$$

If co Y_0 were not pointed, we could represent the origin as a sum of nonzero vectors in co Y_0. This would give a representation of the origin as a sum of nonzero vectors in Y_0, contradicting the pointedness of Y_0. Thus co Y_0 is pointed.

Proving that co Y_0 is closed when Y_0 is pointed amounts to proving in the case of the closed cone

$$W = Y \times \cdots \times Y \subset (R^m)^{m+1}$$

and linear transformation

$$A : w = (y^1, \ldots, y^{m+1}) \to y^1 + \cdots + y^{m+1}$$

that

$$\text{if } w \in W \text{ and } A(w) = 0 \text{ imply } w = 0, \quad \text{then } A(W) \text{ is closed.} \tag{6.36}$$

Suppose $A(W)$ were not closed. Then there would exist $w^j \in W$ such that $A(w^j) \to q \notin A(W)$. The sequence $\{w^j\}$ could not have a bounded subsequence, for if so it would have a cluster point w, and then $A(w) = q$. Therefore $|w^j| \to \infty$, and for $\bar{w}^j = w^j/|w^j|$ we would have $\bar{w}^j \in W$ (because W is a cone) and

$$A(\bar{w}^j) = A(w^j)/|w^j| \to 0.$$

Since $|\bar{w}^j| = 1$ and W is closed, the sequence $\{\bar{w}^j\}$ would have a cluster point $\bar{w} \in W$ satisfying $|\bar{w}| = 1$ and $A(\bar{w}) = 0$. This argument verifies (6.36) and finishes the proof of Proposition 15.

Remark. The need for some further conditions on Y_0 in order to ensure equality in (6.30) is demonstrated by

$$Y = \{(y_1, y_2) \in R^2 \mid y_1^2 \le |y_2|\}, \quad Y_0 = \{(y_1, y_2) \in R^2 \mid y_1 = 0\}.$$

In this case one has (6.28) satisfied but $\operatorname{cl} \operatorname{co} N = \{(y_1, y_2, \eta) \mid \eta \le 0\}$, so that

$$\{(y_1, y_2) \mid (y_1, y_2, 0) \in \operatorname{cl} \operatorname{co} N\} = R^2, \quad \operatorname{cl} \operatorname{co} Y_0 = Y_0 = R^1 \times \{0\}.$$

7. Application to generalized directional derivatives

The estimates in Theorem 2 lead to results about the various derivative functions $p^{\uparrow}$, p^0, p^+, p_+ and p' discussed in Section 2. We have already seen one consequence in Corollary 5; there (2.10) holds, and in particular $p'(u, v; h, k) = y \cdot h + z \cdot k$ for all (h, k).

Theorem 3. *With (u, v) and X satisfying the hypothesis of Theorem 2, let (h, k) be a vector belonging to the closed convex cone*

$$G = \bigcap_{x \in X} \{(h, k) \mid y \cdot h + z \cdot k \le 0 \text{ for all } (y, z) \in K_0(u, v, x)\}. \tag{7.1}$$

If either $p^{\uparrow}(u, v; h, k) < \infty$ or there is at least one $x \in X$ with $K(u, v, x) \ne \emptyset$, one has

$$p^{\uparrow}(u, v; h, k) \le \sup_{x \in X} \left[\sup_{(y, z) \in K(u, v, x)} \{y \cdot h + z \cdot k\} \right], \tag{7.2}$$

(where an empty supremum is interpreted as $-\infty$). This inequality is valid in particular for all $(h, k) \in \operatorname{int} G$; in fact for such (h, k), (2.9) holds and one has the further estimates

$$p^+(u, v; h, k) \le \inf_{x \in X} \left[\sup_{(y, z) \in K(u, v, x)} \{y \cdot h + z \cdot k\} \right], \tag{7.3}$$

$$p_{+}(u, v; -h, -k) \le \inf_{x\in X}\Big[\sup_{(y,z)\in K(u,v,x)} \{-y\cdot h - z\cdot k\}\Big]. \tag{7.4}$$

Proof. The first estimate (7.2) is obtained from the outer estimate (5.8) in Corollary 1 and the formula

$$p^{\uparrow}(u, v; h, k) = \sup\{y\cdot h + z\cdot k \mid (y, z)\in \partial p(u, v)\},$$

which we know from (2.6) to be correct whenever $p^{\uparrow}(u, v; h, k) < \infty$ or $\partial p(u, v) \neq \emptyset$. In taking the supremum of $y\cdot h + z\cdot k$ over all (y, z) belonging to the right side of (5.8), we can certainly ignore the 'cl co.' Thus for the sets

$$M = \bigcup_{x\in X} K(u, v, x), \qquad M_0 = \bigcup_{x\in X} K_0(u, v, x), \tag{7.5}$$

we have

$$p^{\uparrow}(u, v; h, k) \le \sup(y\cdot h + z\cdot k + y^0\cdot h + z^0\cdot k \mid (y, z)\in M,\ (y^0, z^0)\in M_0\}$$

provided that either $p^{\uparrow}(u, v; h, k) < \infty$ or $M + M_0 \neq \emptyset$. Here M_0 is actually a cone (not necessarily convex) which contains $(0, 0)$, and G is its polar, so that

$$\sup\{y^0\cdot h + z^0\cdot k \mid (y^0, z^0)\in M_0\} = \begin{cases} 0, & \text{if } (h, k)\in G,\\ \infty, & \text{if } (h, k)\notin G.\end{cases}$$

Thus $M + M_0 \neq \emptyset$ if and only if $M \neq \emptyset$, and for $(h, k)\in G$ the right side of (7.6) reduces to

$$\sup\{y\cdot h + z\cdot k \mid (y, z)\in M\}.$$

In this manner one obtains the validity of (7.2) for all cases having either $p^{\uparrow}(u, v; h, k) < \infty$ or $M \neq \emptyset$, as asserted.

We have already noted in Corollary 3 that p is directionally Lipschitzian with respect to $(h, k)\in \operatorname{int} G$, which means that (2.9) holds (see Section 2).

To derive (7.3), we initially fix any $x\in X$ and consider the modified problem $(\tilde{P}_{u,v})$ in Proposition 11. As long as ϵ is small enough, this has x as its unique optimal solution and again satisfies all our assumptions, including tameness (with respect to $\tilde{A} = A\cup\{x\}$). The results obtained so far for $(P_{u,v})$ can therefore be applied to $(\tilde{P}_{u,v})$ with $\tilde{X} = \{x\}$: for (h, k) belonging to the interior of

$$\tilde{G} = \{(h, k) \mid y\cdot h + z\cdot k \le 0 \text{ for all } (y, z)\in K_0(u, v, x)\}, \tag{7.7}$$

one has

$$\tilde{p}^{\uparrow}(u, v; h, k) \le \sup_{(y,z)\in K(u,v,x)} \{y\cdot h + k\cdot z\}, \tag{7.8}$$

and moreover (2.9) holds for $\tilde{p}$, so that actually

$$\tilde{p}^{+}(u, v; h, k) \le \tilde{p}^{\uparrow}(u, v; h, k). \tag{7.9}$$

At the same time we have $\tilde{p}$ and p related by (6.3) in Proposition 11, and this implies

$$p^+(u, v; h, k) \le \tilde{p}^+(u, v; h, k). \tag{7.10}$$

Putting together (7.8), (7.9) and (7.10), we see that

$$p^+(u, v; h, k) \le \sup_{(y, z) \in K(u, v, x)} \{y \cdot h + z \cdot k\} \quad \text{when } (h, k) \in \operatorname{int} \tilde{G}.$$

Since $\tilde{G} \supset G$, and x was an arbitrary point of X, the truth of (7.3) for all $(h, k) \in \operatorname{int} G$ is immediate from this.

The argument for (7.4) is different. Bear in mind that p is finite and lower semicontinuous at (u, v) under our hypothesis (cf. Proposition 8). Denoting the right side of (7.2) by β, we see that (7.4) can be written in the form

$$\limsup_{\substack{(h', k') \to (h, k) \\ t \downarrow 0}} \frac{p(u, v) - p(u - th', v - tk')}{t} \le \beta, \tag{7.11}$$

while what we know from (7.2) and (2.10) is that

$$\limsup_{\substack{(u', v') \to_p (u, v) \\ (h', k') \to (h, k) \\ t \uparrow 0}} \frac{p(u' + th', v' + tk') - p(u', v')}{t} \le \beta. \tag{7.12}$$

Our task will be to derive (7.11) from (7.12). Let α denote the value of the 'lim sup' in (7.11), and consider any consequences $(h^j, k^j) \to (h, k)$ and $t_j \downarrow 0$ for which it is attained:

$$\lim_{j \to \infty} \frac{p(u, v) - p(u - t_j h^j, v - t_j k^j)}{t} = \alpha. \tag{7.13}$$

We need to show $\alpha \le \beta$, and for this purpose it is enough to look at the case where $\alpha > -\infty$. Passing to subsequences in (7.13) if necessary, we can suppose that

$$\gamma = \lim_{j \to \infty} p(u - t_j h^j, v - t_j k^j)$$

exists. Since $\alpha > -\infty$ and $t_j > 0$ in (7.13) it must be true that $\gamma \le p(u, v)$, yet the opposite inequality must hold too, because p is lower semicontinuous at (u, v). Hence

$$p(u - t_j h^j, v - t_j k^j) \to p(u, v). \tag{7.14}$$

Define $(u^j, v^j) = (u - t_j h^j, v - t_j k^j)$. Then

$$\lim_{j \to \infty} \frac{p(u^j + t_j h^j, v^j + t_j h^j) - p(u^j, v^j)}{t_j} = \alpha$$

by (7.13), and $(u^j, v^j) \to_p (u, v)$ by (7.14). It follows from (7.12) that $\alpha \le \beta$, and this completes our proof.

Remark 1. Inequality (7.4) could also be expressed as

$$p^{-}(u, v; h, k) \leq \sup_{x \in X} \Big[\sup_{(y, z) \in K(u, v, x)} \{y \cdot h + z \cdot k\} \Big], \tag{7.15}$$

where in parallel to the definition (2.11) of $p^{+}(u, v; h, k)$ one takes

$$p^{-}(u, v; h, k) = \limsup_{\substack{(h', k') \to (h, k) \\ t \uparrow 0}} \frac{p(u + th', v + tk') - p(u, v)}{t}. \tag{7.16}$$

Remark 2. Inequality (7.3) holds in a more general form, as shown by the proof: for an *arbitrary* set X_0 of optimal solutions to $(P_{u,v})$, if $(h, k) \in \operatorname{int} G_0$, where G_0 is the cone obtained when X is replaced by X_0 in (7.1), then (7.3) too is valid with X replaced by X_0.

Corollary 1. *With (u, v) and X satisfying the hypothesis of Theorem 2, suppose that the constraint qualification $K_0(u, v, x) = \{(0, 0)\}$ holds for every $x \in X$. Then for all $(h, k) \in \mathbb{R}^m \times \mathbb{R}^d$ one has (2.9) and*

$$p^{+}(u, v; h, k) \leq \inf_{x \in X} \Big[\sup_{(y, z) \in K(u, v, x)} \{y \cdot h + z \cdot k\} \Big], \tag{7.17}$$

$$p_{+}(u, v; h, k) \geq \inf_{x \in X} \Big[\inf_{(y, z) \in K(u, v, x)} \{y \cdot h + z \cdot k\} \Big]. \tag{7.18}$$

If in addition $K(u, v, x)$ is a singleton $\{(y(x), z(x))\}$ for each $x \in X$, then the derivatives $p'(u, v; h, k)$ exist, and in fact

$$p^{+}(u, v; h, k) = p_{+}(u, v; h, k) = \inf_{x \in X} \{y(x) \cdot h + z(x) \cdot k\}. \tag{7.19}$$

Proof. This is the case of Theorem 3 where G is all of $\mathbb{R}^m \times \mathbb{R}^d$, so that (h, k) and $(-h, -k)$ both always belong to int G. The inner 'sup' in (7.17) and 'inf' in (7.18) coincide, of course, when $K(u, v, x)$ is a singleton.

Corollary 1 generalizes results of Gauvin and Tolle [13], Gauvin [11], Gauvin and Dubeau [12] in the smooth case (a) of $(P_{u,v})$, and of Auslender [2] in the somewhat more general case where the *inequality* constraints need not be smooth. Corollary 1 allows nonsmooth equality constraints too, plus abstract constraints represented by $(v, x) \in D$, and at the same time yields stronger conclusions in terms of Hadamard derivatives instead of just Dini derivatives.

Corollary 2. *With (u, v) and X as in the hypothesis of Theorem 2, and G the cone in (7.1), if* $\operatorname{int} G \neq \emptyset$ *one has either $p^{+}(u, v; h, k) > -\infty$ for all $(h, k) \in \operatorname{int} G$ or*

$$p^{+}(u, v; h, k) = p_{+}(u, v; h, k) = -\infty \quad \text{for all } (h, k) \in \operatorname{int} G, \tag{7.20}$$

the latter case occurring if and only if $K(u, v, x) = \emptyset$ for some $x \in X$.

Proof. Apply (7.3) and use the fact that $p_+ \le p^+$.

Corollary 3. *In the case where no parameter vector v (or corresponding multiplier vector z) is being considered, and all the explicit constraints in* (P_u) *are inequalities (notationally: $s = m$), suppose u is such that* (P_u) *is tame, and let X be the set of all optimal solutions to* (P_u). *Then for every strictly negative vector h (i.e., $-h \in \operatorname{int} \mathbb{R}^m_+$) one has*

$$p^+(u; h) \le \inf_{x \in X} \Big[\sup_{y \in K(u,x)} y \cdot h \Big] \le 0. \tag{7.21}$$

Proof. Here $K_0(u, x) \subset \mathbb{R}^m_+$ for all $x \in X$, so that the cone G in (7.1) includes $-\mathbb{R}^m_+$. Every strictly negative h therefore belongs to $\operatorname{int} G$, and (7.21) can be obtained as a special case of (7.3).

Our final result treats only a special, but nevertheless very important class of problems. It extends the marginal value theorem of Gol'shtein [5, Section 7] to the case where the set of Kuhn–Tucker pairs associated with $(P_{u,v})$ is not necessarily compact. Again, conclusions are obtained for Hadamard derivatives rather than just Dini derivatives.

Theorem 4. *In the mixed smooth-convex case* (c) *in Section* 1, *and with* (u, v) *and X such that the hypothesis of Theorem* 2 *is satisfied, one has for all* (h, k):

$$p_+(u, v; h, k) \ge \inf_{x \in X} \Big[\sup_{(y,z) \in K(u,v,x)} \{ y \cdot h + z \cdot k \} \Big]. \tag{7.22}$$

The set of vectors $y \in \mathbb{R}^m$ satisfying for a given $x \in X$ the complementary slackness conditions (4.3) *and*

$$0 \in \sum_{i=1}^{m} y_i \partial_x f_i(v, x) + N_C(x),$$

is actually a closed convex cone Y_0 independent of x. The convex cone in (7.1) *takes the form*

$$G = \bigcap_{x \in X} \Big\{ (h, k) \,\Big|\, \sum_{i=1}^{m} y_i [\nabla_v f_i(x, v) k + h_i] \le 0 \text{ for all } y \in Y_0 \Big\}, \tag{7.23}$$

and for all $(h, k) \in \operatorname{int} G$ one has

$$p^+(u, v; h, k) = p_+(u, v; h, k) = \min_{x \in X} \Big[\sup_{(y,z) \in K(u,v,x)} \{ y \cdot h + z \cdot k \} \Big]. \tag{7.24}$$

Proof. Since (7.22) is trivial if $p_+(u, v; h, k) = +\infty$, we can suppose in proving (7.22) that

$$\infty > p_+(u, v; h, k) = \lim_{j \to \infty} \frac{p(u + t_j h^j, v + t_j k^j) - p(u, v)}{t_j}. \tag{7.25}$$

for certain sequences $t_j \downarrow 0$ and $(h^j, k^h) \to (h, k)$. Let

$$(u^j, v^j) = (u + t_j h^j, v + t_j k^j) \to (u, v).$$

Then $p(u^j, v^j) \to p(u, v)$ by (7.25) and the lower semicontinuity of p at (u, v), the latter being a consequence of the tameness condition in the hypothesis of Theorem 2 (cf. Proposition 8). Thus $(u^j, v^j) \to_p (u, v)$. Introduce next a mapping ξ as in Proposition 10 whose cluster points as $(u', v') \to_p (u, v)$ all belong to the set A invoked in the tameness definition (5.1). Setting $x^j = \xi(u^j, v^j)$ and passing to subsequences of necessary, we get a convergent sequence of optimal solutions x^j to (P_{u^j, v^j}) whose limit is a certain optimal solution $x \in A$ to $(P_{u,v})$. Then $x \in X$, since under the hypothesis of Theorem 2 every optimal solution to $(P_{u,v})$ in A is also in X. Note that $f_0(v^j, x^j) = p(u^j, v^j)$ and $f_0(v, x) = p(u, v)$, and hence

$$\lim_{j \to \infty} \frac{f_0(v^j, x^j) - f_0(v, x)}{t_j} = p_+(u, v; h, k) \tag{7.26}$$

by (7.25). Consider now an arbitrary $(y, z) \in K(u, v, x)$. This satisfies (4.3) and (4.4), but since we are dealing with the mixed case (b) of $(P_{u,v})$, (4.4) can be written as (4.11), or in terms of the function

$$l = f_0 + \sum_{i=1}^{m} y_i f_i$$

even more simply as

$$0 \in \partial_x l(v, x) + N_C(x) \quad \text{and} \quad z = \nabla_v l(v, x). \tag{7.27}$$

Here l inherits from the functions f_i the property of being convex in the second argument, and $\mathscr{C}^1$ in the first argument with gradient depending continuously on *both* arguments. This joint continuity ensures (via the mean value theorem) that actually

$$\lim_{j \to \infty} \frac{l(v + t_j k^j, x^j) - l(v, x^j)}{t_j} = \nabla_v l(v, x) \cdot k = z \cdot k, \tag{7.28}$$

a fact that will be put to use presently. The convexity property of l, on the other hand, allows us to read the first condition in (7.27) as saying that $l(v, \cdot)$ attains it minimum over C at x (cf. [26, Theorem 27.4]). Since x^j is feasible for (P_{u^j, v^j}), and hence in particular $x^j \in C$, it follows from this that

$$l(v, x^j) \geq l(v, x) \tag{7.29}$$

and from (4.3) that

$$f_0(v, x) = f_0(v, x) + \sum_{i=1}^{m} y_i[f_i(v, x) + u_i] = l(v, x) + y \cdot u, \tag{7.30}$$

$$f_0(v^j, x^j) \geq f_0(v^j, x^j) + \sum_{i=1}^{m} y_i[f_i(v^j, x^j) + u_i^j] = l(v^j, x^j) + y \cdot u^j. \tag{7.31}$$

Therefore

$$\begin{aligned} f_0(v^j, x^j) - f_0(v, x) &\geq [l(v^j, x^j) + y \cdot u^j] - [l(u, x) + y \cdot u] \\ &\geq y \cdot (u^j - u) + l(v^j, x^j) - l(v, x^j) \\ &= t_j y \cdot h^j + [l(v + t_j k^j, x^j) - l(v, x^j)]. \end{aligned}$$

Using this estimate in (7.26) and invoking (7.28), we get

$$p_+(u, v; h, k) \geq y \cdot h + z \cdot k.$$

This being true for arbitrary $(y, z) \in K(u, v, x)$, we conclude that

$$p_+(u, v; h, k) \geq \sup_{(y, z) \in K(u, v, x)} \{y \cdot h + z \cdot k\} \tag{7.32}$$

for the particular $x \in X$ which has been constructed, and hence that (7.22) is indeed true.

The rest of the proof of Theorem 4 is mostly a matter of applying Theorem 3, specifically (7.3). The special form for G is readily derived from the fact that condition (4.5) in the definition of $K_0(u, v, x)$ reduces in the present case to (4.12) (cf. also Proposition 6). As pointed out in Section 4 in the remarks following (4.12), the first condition in (4.12), together with (4.3) and the feasibility of x, constitute a certain saddlepoint condition on (x, y). As is well known, the set of saddlepoints of a given function is always a product set; the set of y's corresponding to a given x is independent of the choice of x.

Corollary. *Under the assumptions in Theorem* 4, *if the set* Y_0 *consists of just* $y = 0$, *then* p *is locally Lipschitzian around* (u, v) *and* (7.24) *holds for all* (h, k).

Proof. To say that $Y_0 = \{0\}$ is to say that $K_0(u, v, x) = \{(0, 0)\}$ for each $x \in X$. Then p is locally Lipschitzian by Corollary 2 of Theorem 2 in Section 5.

References

[1] J.P. Aubin, "Further properties of Lagrange multipliers in nonsmooth optimization", *Applied Mathematics and Optimization* 6 (1980) 79–90.

[2] A. Auslender, "Differential stability in nonconvex and non differentiable programming", in: P. Huard, ed., *Point-to-set maps and mathematical programming,* Mathematical Programming Studies, Vol. 10 (North-Holland, Amsterdam, 1979) pp. 29–41.

[3] F.H. Clarke, "A new approach to Lagrange Multipliers", *Mathematics of Operations Research* 1 (1976) 165–174.
[4] F.H. Clarke, "Generalized gradients and applications", *Transactions of the American Mathematical Society* 205 (1975) 247–262.
[5] F.H. Clarke, "Generalized gradients of Lipschitz functionals", *Advances in Mathematics* (1980).
[6] F.H. Clarke and J.P. Aubin, "Shadow prices and duality for a class of optimal control problems", *SIAM Journal on Control and Optimization* 17 (1979) 567–586.
[7] V.E. Dem'janov and V.N. Malozemov, "The theory of nonlinear minimal problems", *Russian Mathematical Surveys* 26 (1971) 57–115.
[8] V.E. Dem'janov and A.B. Pevnyi, "First and second marginal values of mathematical programming problems", *Soviet Mathematics Doklady* 13 (1972) 1502–1506.
[9] J.P. Evans and F.J. Gould, "Stability in nonlinear programming", *Operations Research* 18 (1970) 107–118.
[10] A. Fiacco and W.P. Hutzler, "Extension of the Gauvin–Tolle optimal value differential stability results to general mathematical programs", *Mathematical Programming Study*, to appear.
[11] J. Gauvin, "The generalized gradients of a marginal function in mathematical programming", *Mathematics of Operations Research* 4 (1979) 458–463.
[12] J. Gauvin and F. Dubeau, "Differential properties of the marginal function in mathematical programming", *Mathematical Programming Study*, to appear.
[13] J. Gauvin and J.W. Tolle, "Differential stability in nonlinear programming", *SIAM Journal on Control and Optimization* 15 (1977) 294–311.
[14] B. Gollan, "A general perturbation theory for abstract optimization problems", *Journal of Optimizations Theory and Applications*, to appear.
[15] E.G. Gol'shtein, *Theory of convex programming*, Translations of Mathematical Monographs 36 (American Mathematical Society, Providence RI, 1972).
[16] H.J. Greenberg and W.P. Pierskalla, "Extensions of the Evans–Gould stability results for mathematical programs", *Operations Research* 20 (1972) 143–153.
[17] J.-B. Hiriart-Urruty, "Gradients généralisés de fonctions marginales", *SIAM Journal on Control and Optimization* 16 (1978) 301–316.
[18] J.-B. Hiriart-Urruty, "Refinements of necessary optimality conditions in nondifferentiable programming I", *Applied Mathematics and Optimization* 5 (1979) 63–82.
[19] J.-B. Hiriart-Urruty, "Refinements of necessary optimality conditions in nondifferentiable programming II", to appear.
[20] W.W. Hogan, "Directional derivatives for extremal-value functions with applications to the completely convex case", *Operations Research* 21 (1973) 188–209.
[21] F. Lempio and H. Maurer, "Differential stability in infinite dimensional nonlinear programming", *Applied Mathematics and Optimization*, to appear.
[22] E.S. Levitin, "On the local perturbation theory of mathematical programming in a Banach space", *Soviet Mathematics Doklady* 16 (1975).
[23] E.S. Levitin, "On the perturbation theory of nonsmooth extremal problems with constraints", *Soviet Mathematics Doklady* 16 (1975).
[24] O.L. Mangasarian and S. Fromovitz, "The Fritz John necessary optimality conditions in the presence of equality and inequality constraints", *Journal of Mathematical Analysis and Applications* 17 (1967) 37–47.
[25] R.T. Rockafellar, "Clarke's tangent cones and the boundaries of closed sets in $\mathbb{R}^n$". *Nonlinear Analysis, Theory, Methods and Applications* 3 (1979) 145–154.
[26] R.T. Rockafellar, Convex Analysis (Princeton University Press, Princeton, NJ, 1970).
[27] R.T. Rockafellar, "Directionally Lipschitzian functions and subdifferential calculus", *Proceedings of the London Mathematical Society* 39 (1979) 331–355.
[28] R.T. Rockafellar, "Generalized directional derivatives and subgradients of nonconvex functions", *Canadian Journal of Mathematics* 32 (1980) 257–280.
[29] R.T. Rockafellar, "Proximal subgradients, marginal values and augmented Lagrangians in nonconvex optimization", *Mathematics of Operations Research* 6 (1981).

[30] R.T. Rockafellar, *La théorie des sous-gradients et ses applications: Fonctions convexes et non convexes*, Collection Chaire Aisenstadt (Presses de l'Université de Montréal, Montréal, Québec, 1979).

[31] S.M. Robinson, "Stability theory for systems of inequalities, part I: Linear systems", *SIAM Journal on Numerical Analysis* 12 (1976) 754–769.

[32] S.M. Robinson, "Stability theory for systems of inequalities, part II: Differential nonlinear systems", *SIAM Journal on Numerical Analysis* 13 (1976) 497–513.

[33] A.C. Williams, "Marginal values in linear programming", *SIAM Journal on Applied Mathematics* 11 (1963) 82–99.

Mathematical Programming Study 17 (1982) 67–76.
North-Holland Publishing Company

A MODEL ALGORITHM FOR COMPOSITE NONDIFFERENTIABLE OPTIMIZATION PROBLEMS

R. FLETCHER

Mathematics Department, University of Dundee, Dundee DD1 4HN, Scotland, UK

Received 29 May 1980
Revised manuscript received 1 June 1981

Composite functions $\phi(x) = f(x) + h(c(x))$, where f and c are smooth and h is convex, encompass many nondifferentiable optimization problems of interest including exact penalty functions in nonlinear programming, nonlinear min–max problems, best nonlinear L_1, L_2 and L_∞ approximation and finding feasible points of nonlinear inequalities. The idea is used of making a linear approximation to $c(x)$ whilst including second order terms in a quadratic approximation to $f(x)$. This is used to determine a composite function ψ which approximates $\phi(x)$ and a basic algorithm is proposed in which ψ is minimized on each iteration. If the technique of a step restriction (or trust region) is incorporated into the algorithm, then it is shown that global convergence can be proved. It is also described briefly how the above approximations ensure that a second order rate of convergence is achieved by the basic algorithm.

Key words: Composite Functions, Nondifferentiable Optimization, Approximations, Rate of Convergence.

1. Introduction

Most nondifferentiable optimization (NDO) problems can be expressed in the form of minimizing a *composite function*

$$\underset{x\in\mathbb{R}^n}{\text{minimize}}\ \phi(x) \overset{\Delta}{=} f(x) + h(c(x)) \tag{1.1}$$

where $f(x)$ ($\mathbb{R}^n \to \mathbb{R}$) and $c(x)$ ($\mathbb{R}^n \to \mathbb{R}^m$) are smooth functions ($\mathbb{C}^1$ at least) and $h(c)$ ($\mathbb{R}^m \to \mathbb{R}$) is convex. An example of this is an exact penalty function for non-linear programming in which $f(x)$ is the weighted objective function and $c(x)$ is the vector of constraint functions in the problem. For equality constraints $c(x) = 0$, $h(c)$ is defined as $\|c\|$ and for inequality constraints $c(x) \le 0$, $h(c)$ is defined as $\|c^+\|$ where $c_i^+ = \max(c_i, 0)$. The function $\|\cdot^+\|$ is convex when the norm is monotonic ($|x| \le |y| \Rightarrow \|x\| \le \|y\|$) which includes all L_p and scaled L_p norms for $1 \le p \le \infty$. In practice the $\|\cdot\|_1$ currently attracts the most interest. Other NDO problems arise in which $f(x)$ is not present ($f = 0$). These include finite min–max problems ($h(c) \overset{\Delta}{=} \max_i c_i$), best L_1, L_2 or L_∞ approximation problems ($h(c) \overset{\Delta}{=} \|c\|_p$ $p = 1, 2, \infty$) (which includes the solution of systems of nonlinear equations when $m = n$) and problems of finding feasible points of nonlinear systems of in-equalities ($h(c) \overset{\Delta}{=} \|c^+\|$). It is convenient to treat all these as special cases of (1.1).

In many of these examples $h(c)$ is a *polyhedral convex function*

$$h(\mathbf{c}) = \max_i(\mathbf{c}^T\mathbf{h}_i + b_i) \tag{1.2}$$

made up of a finite number of supporting hyperplanes, where the $\mathbf{h}_i$ and b_i are given. In all cases of interest here $b_i = 0\ \forall i$, and the vectors $\mathbf{h}_i$ (given as columns of a matrix $\mathbf{H}$) are

$h(\mathbf{c}) = \max_i c_i,$	$\mathbf{H} = \mathbf{I}$	$(m \times m),$
$h(\mathbf{c}) = \Vert\mathbf{c}^+\Vert_\infty$	$\mathbf{H} = [\mathbf{I}, \mathbf{0}]$	$(m \times (m+1)),$
$h(\mathbf{c}) = \Vert\mathbf{c}\Vert_\infty$	$\mathbf{H} = [\mathbf{I}, -\mathbf{I}]$	$(m \times 2m),$
$h(\mathbf{c}) = \Vert\mathbf{c}^+\Vert_1,$	columns of $\mathbf{H}$ are all possible combinations of 1 and 0 $(m \times 2^m)$,	
$h(\mathbf{c}) = \Vert\mathbf{c}\Vert_1,$	columns of $\mathbf{H}$ are all possible combinations of 1 and -1 $(m \times 2^m)$.	

An example of the last case when $m = 2$ is the matrix

$$\mathbf{H} = \begin{bmatrix} 1 & 1 & -1 & -1 \\ 1 & -1 & 1 & -1 \end{bmatrix}.$$

First order conditions for the solution of (1.1) at a point $\mathbf{x}^*$ are that $\mathbf{0} \in \partial\phi^*$ or equivalently that

$$\max_{\mathbf{g} \in \partial\phi^*} \mathbf{s}^T\mathbf{g} \geq 0, \quad \forall \mathbf{s} \tag{1.3}$$

where $\partial\phi^*(=\partial\phi(\mathbf{x}^*))$ is the generalized gradient of ϕ at $\mathbf{x}^*$ [2] defined by

$$\partial\phi(\mathbf{x}) = \{\mathbf{g} : \mathbf{g} = \nabla f + \mathbf{A}\boldsymbol{\lambda}, \forall \boldsymbol{\lambda} \in \partial h\} \tag{1.4}$$

where $\mathbf{A}$ is the Jacobian matrix with columns ∇c_i and ∂h is the subdifferential of $h(\mathbf{c})$ at $\mathbf{c}(=\mathbf{c}(\mathbf{x}))$. The left hand side of (1.3) can be interpreted as the directional derivative of ϕ along $\mathbf{s}$ and (1.3) states that this is nonnegative for all $\mathbf{s}$, which is a necessary condition for a local solution. By virtue of (1.4), alternative forms of (1.3) are that there exists $\boldsymbol{\lambda}^* \in \partial h^*$ such that

$$\mathbf{g}^* + \mathbf{A}^*\boldsymbol{\lambda}^* = \mathbf{0} \tag{1.5}$$

or that

$$\max_{\boldsymbol{\lambda} \in \partial h^*} \mathbf{s}^T(\mathbf{g}^* + \mathbf{A}^*\boldsymbol{\lambda}) \geq 0, \quad \forall \mathbf{s} \tag{1.6}$$

where $\mathbf{g}^* = \nabla f(\mathbf{x}^*)$. The parameters $\boldsymbol{\lambda}^*$ are closely related to Lagrange multipliers in a constrained minimization problem. For a polyhedral convex function the subdifferential has the simple form

$$\partial h(\mathbf{c}) = \operatorname*{conv}_{i \in \mathcal{A}} \mathbf{h}_i \tag{1.7}$$

where

$$\mathcal{A} = \{i: h(c) = c^{\mathrm{T}}h_i + b_i\} \tag{1.8}$$

is the set of indices at which the max is achieved in (1.2). These results are given for example by Fletcher [6] where the material of this paper is developed in more detail and with more background. The case that $h(c) \overset{\Delta}{=} \|c\|$ is considered by Fletcher and Watson [7].

2. A model algorithm

If h is polyhedral, it follows from (1.2) that the solution to (1.1) is equivalent to solving the nonlinear programming problem

$$\begin{aligned} &\underset{x,v}{\text{minimize}} \quad v, \\ &\text{subject to} \quad v - f(x) - c(x)^{\mathrm{T}}h_i \geq b_i, \quad \forall i. \end{aligned} \tag{2.1}$$

This can be solved by any convenient algorithm for nonlinear programming. However an obvious choice is the well-known SOLVER method (e.g. [3]) based on iterated quadratic programming. This idea is well known, especially in the Russian literature, for example [12]. Another interesting algorithm for solving (1.1) is based on making a linear Taylor series approximation $l^{(k)}(\delta)$ to $c(\delta)$ about a current iterate $x^{(k)}$. This is defined by

$$c(x^{(k)} + \delta) \simeq l^{(k)}(\delta) \overset{\Delta}{=} c^{(k)} + A^{(k)\mathrm{T}}\delta \tag{2.2}$$

($c^{(k)} = c(x^{(k)})$ etc.). Also a quadratic approximation $q^{(k)}(\delta)$ to $f(x)$ is defined by

$$f(x^{(k)} + \delta) \simeq q^{(k)}(\delta) \overset{\Delta}{=} f^{(k)} + g^{(k)\mathrm{T}}\delta + \tfrac{1}{2}\delta^{\mathrm{T}}W^{(k)}\delta \tag{2.3}$$

where $W^{(k)} = \nabla^2 f^{(k)} + \sum_i \lambda_i^{(k)}\nabla^2 c_i^{(k)}$. This is a Taylor series for $f(x)$ about $x^{(k)}$ together with additional terms in the Hessian which account for curvature in the functions $c_i(x)$. The parameters $\lambda_i^{(k)}$ are estimates of the optimum parameters λ^* in (1.5). These functions are used analogously to f and c in (1.1) to define a composite function

$$\psi^{(k)}(\delta) = q^{(k)}(\delta) + h(l^{(k)}(\delta)) \tag{2.4}$$

which therefore approximates $\phi(x^{(k)} + \delta)$. The basic algorithm therefore is to choose $\delta^{(k)}$ to minimize $\psi^{(k)}(\delta)$, and to set $x^{(k+1)} = x^{(k)} + \delta^{(k)}$ and $\lambda^{(k+1)}$ as the λ parameters at the solution to (2.4). It is convenient to refer to this as the *QL method* since the $\mathbb{C}^1$ functions which occur in (1.1) are approximated by a quadratic and a linear function respectively. It is shown in the Appendix that the QL method is equivalent to the SOLVER method applied to (2.1).

The basic technique for minimizing $\phi(x)$ is likely to have a second order rate of convergence (by analogy with the SOLVER method—see Section 4) but is not likely to be globally convergent. (For $m = 0$ the term in $h(\cdot)$ is ignored and the method is just Newton's method for minimizing $f(\boldsymbol{x})$). A successful modification for inducing global convergence is the restricted step or trust region approach (e.g. [4]) in which a step restriction $\|\boldsymbol{\delta}\| \le h^{(k)}$ is enforced at each iteration which aims to measure the region in which the Taylor series approximations (2.2) and (2.3) are valid. Notationally the parameter $h^{(k)}$ should not be confused with the function $h(\boldsymbol{c})$. Thus the subproblem which is solved on iteration k is

$$\begin{aligned} &\underset{\boldsymbol{\delta}}{\text{minimize}} && \psi^{(k)}(\boldsymbol{\delta}), \\ &\text{subject to} && \|\boldsymbol{\delta}\| \le h^{(k)}. \end{aligned} \tag{2.5}$$

The radius $h^{(k)}$ of the trust region is adjusted adaptively to be as large as possible subject to reasonable agreement between $\phi(\boldsymbol{x}^{(k)} + \boldsymbol{\delta})$ and $\psi^{(k)}(\boldsymbol{\delta})$ being maintained. This can be quantified by defining the *actual reduction*

$$\Delta\phi^{(k)} = \phi^{(k)} - \phi(\boldsymbol{x}^{(k)} + \boldsymbol{\delta}^{(k)}) \tag{2.6}$$

and the *predicted reduction*

$$\Delta\psi^{(k)} = \phi^{(k)} - \psi^{(k)}(\boldsymbol{\delta}^{(k)}). \tag{2.7}$$

Then the ratio

$$r^{(k)} = \Delta\phi^{(k)}/\Delta\psi^{(k)} \tag{2.8}$$

measures the extent to which ϕ and $\psi^{(k)}$ agree in a neighborhood of $\boldsymbol{x}^{(k)}$. The choice of norm in (2.5) is arbitrary but is most likely to be the $\|\cdot\|_2$ or $\|\cdot\|_\infty$ in practice. A model algorithm in which these ideas are used is the following (the *k-th* iteration is described).

(i) Given $\boldsymbol{x}^{(k)}$, $\boldsymbol{\lambda}^{(k)}$ and $h^{(k)}$, calculate $f^{(k)}$, $\boldsymbol{g}^{(k)}$, $\boldsymbol{c}^{(k)}$, $\boldsymbol{A}^{(k)}$ and $\boldsymbol{W}^{(k)}$ which determines $\phi(\boldsymbol{x}^{(k)})$ and $\psi^{(k)}(\boldsymbol{\delta})$.

(ii) Find a global solution $\boldsymbol{\delta}^{(k)}$ to (2.5)

(iii) Evaluate $\phi(\boldsymbol{x}^{(k)} + \boldsymbol{\delta}^{(k)})$ and hence $\Delta\phi^{(k)}$, $\Delta\psi^{(k)}$ and $r^{(k)}$.

(iv) If $r^{(k)} < 0.25$ set $h^{(k+1)} = \|\boldsymbol{\delta}^{(k)}\|/4$; if $r^{(k)} > 0.75$ and $\|\boldsymbol{\delta}^{(k)}\| = h^{(k)}$ set $h^{(k+1)} = 2h^{(k)}$; otherwise set $h^{(k+1)} = h^{(k)}$. (2.9)

(v) If $r^{(k)} \le 0$ set $\boldsymbol{x}^{(k+1)} = \boldsymbol{x}^{(k)}$, $\boldsymbol{\lambda}^{(k+1)} = \boldsymbol{\lambda}^{(k)}$; else $\boldsymbol{x}^{(k+1)} = \boldsymbol{x}^{(k)} + \boldsymbol{\delta}^{(k)}$, $\boldsymbol{\lambda}^{(k+1)} =$ multipliers from solving (2.5).

The motivation for changing the parameter $r^{(k)}$ is that if poor agreement occurs, then the Taylor series is not valid for such large $h^{(k)}$ and so the trust region is reduced in size. If good agreement occurs and the trust region is active, then the trust region is unduly restrictive and is therefore increased in size. In practice some rather more sophisticated strategy would be used, especially for reducing $h^{(k)}$, but this need not affect the validity of the proof which follows. The parameters 0.25, 0.75, etc. which arise are arbitrary and are not very sensitive but the values given are typical. The way in which multipliers arise at a solution to (2.5) has not been considered here but is very similar to (2.4) [7]. The only property to emphasize is that

$$\boldsymbol{\lambda}^{(k+1)} \in \partial h(\boldsymbol{l}^{(k)}(\boldsymbol{\delta}^{(k)})) \tag{2.10}$$

which is a uniformly bounded set. (See (1.7) for example: in general if $\boldsymbol{x}^{(k)} \in B$ where B is bounded, then $\boldsymbol{c}^{(k)}$, $\boldsymbol{A}^{(k)}$ and $\boldsymbol{\delta}^{(k)}$, and hence $\boldsymbol{l}^{(k)}(\boldsymbol{\delta}^{(k)})$ are bounded. Furthermore the set $\partial h(\boldsymbol{c})$ is bounded if $\boldsymbol{c}$ is bounded so it follows from (2.10) that the parameters $\boldsymbol{\lambda}^{(k)}$ used in the algorithm are bounded.) Good numerical results with this type of algorithm have been obtained in smooth optimization [4] and in NDO problems which occur in the solution of nonlinear programming problems using an L_1 exact penalty function [5]. However the purpose of this paper is to give convergence results for this type of algorithm.

3. Global convergence

A global convergence result for algorithm (2.9) is presented and requires a preliminary lemma. This relates the directional derivative at $\boldsymbol{x}'$ to the difference quotients taken along a fixed direction s as $\boldsymbol{x}^{(k)} \to \boldsymbol{x}'$.

Lemma 3.1 [2]. *Let S be set of all sequences $\boldsymbol{x}^{(k)} \to \boldsymbol{x}'$, $\epsilon^{(k)} \downarrow 0$, then*

$$\limsup_{S} (\phi(\boldsymbol{x}^{(k)} + \epsilon^{(k)} \mathrm{s}) - \phi^{(k)})/\epsilon^{(k)} = \max_{\mathrm{g} \in \partial \phi'} \mathrm{s}^{\mathrm{T}} \mathbf{g} \tag{3.1}$$

in the sense that the difference quotient is bounded above and the sup of all accumulation points of all sequences in S is the directional derivative.

Proof. Because $\phi(\boldsymbol{x})$ is a locally Lipschitz function [2]; an alternative proof which utilizes the structure of ϕ is given by Fletcher [6].

The main result of this paper relating to algorithm (2.9) from arbitrary initial $\boldsymbol{x}^{(1)}$, $\boldsymbol{\lambda}^{(1)}$ and $h^{(1)}$ (>0) can now be stated.

Theorem 3.1. *Let $\boldsymbol{x}^{(k)} \in B \subset \mathbb{R}^n$ where B is bounded and let f, $\boldsymbol{c}$ be $\mathbb{C}^2$ functions with bounded second derivatives on B. Then there exists an accumulation point $\boldsymbol{x}^\infty$ of algorithm (2.9) at which first order conditions hold, that is*

$$\max_{\mathrm{g} \in \partial \phi^\infty} \mathrm{s}^{\mathrm{T}} \mathbf{g} \geq 0, \quad \forall \mathrm{s}. \tag{3.2}$$

Proof. There exists a convergent subsequence $x^{(k)} \to x^{\infty}$ for which either

(i) $r^{(k)} < 0.25$, $h^{(k+1)} \to 0$ and hence $\|\delta^{(k)}\| \to 0$, or

(ii) $r^{(k)} \geq 0.25$ and $\inf h^{(k)} > 0$.

In either case (3.2) is shown to hold. In case (i) let there exist a descent direction s ($\|s\| = 1$) at x^{∞}, that is

$$\max_{g \in \partial\phi^{\infty}} s^{\mathrm{T}} g = -d, \quad d > 0. \tag{3.3}$$

By Taylor series

$$f(x^{(k)} + \epsilon^{(k)} s) = f^{(k)} + \epsilon^{(k)} s^{\mathrm{T}} g^{(k)} + o(\epsilon^{(k)}) = q^{(k)}(\epsilon^{(k)} s) + o(\epsilon^{(k)}) \tag{3.4}$$

by (2.3), since $\lambda^{(k)}$, $\nabla^2 f^{(k)}$, $\nabla^2 c_i^{(k)}$ and hence $W^{(k)}$ are bounded. Likewise by (2.2)

$$c(x^{(k)} + \epsilon^{(k)} s) = l^{(k)}(\epsilon^{(k)} s) + o(\epsilon^{(k)}) \tag{3.5}$$

and hence by (1.1), the boundedness of ∂h and (2.4) it follows that

$$\begin{aligned} \phi(x^{(k)} + \epsilon^{(k)} s) &= q^{(k)}(\epsilon^{(k)} s) + h(l^{(k)}(\epsilon^{(k)} s) + o(\epsilon^{(k)})) + o(\epsilon^{(k)}) \\ &= q^{(k)}(\epsilon^{(k)} s) + h(l^{(k)}(\epsilon^{(k)} s)) + o(\epsilon^{(k)}) \\ &= \psi^{(k)}(\epsilon^{(k)} s) + o(\epsilon^{(k)}). \end{aligned} \tag{3.6}$$

Writing $\epsilon^{(k)} = \|\delta^{(k)}\|$ and considering a step along s in the subproblem, it follows by the optimality of $\delta^{(k)}$ that

$$\begin{aligned} \Delta\psi^{(k)} &\geq \phi^{(k)} - \psi^{(k)}(\epsilon^{(k)} s) \\ &= \phi^{(k)} - \phi(x^{(k)} + \epsilon^{(k)} s) + o(\epsilon^{(k)}) \\ &\geq \epsilon^{(k)}(d + o(1)) + o(\epsilon^{(k)}) = d\epsilon^{(k)} + o(\epsilon^{(k)}) \end{aligned} \tag{3.7}$$

by (3.1) and (3.3). But (3.6) implies that

$$\Delta\phi^{(k)} = \Delta\psi^{(k)} + o(\epsilon^{(k)})$$

and hence $r^{(k)} = \Delta\phi^{(k)} / \Delta\psi^{(k)} = 1 + o(\epsilon^{(k)}) / \Delta\psi^{(k)} = 1 + o(1)$ from (3.7) since $d > 0$, which contradicts $r^{(k)} < 0.25$. Thus $d \leq 0$ for all s and hence (3.2) holds at x^{∞}.

In case (ii), $\phi^{(1)} - \phi^{\infty} \geq \sum_k \Delta\phi^{(k)}$ summed over the subsequence, so from (2.8) $r^{(k)} \geq 0.25$ implies that $\Delta\psi^{(k)} \to 0$ since $\phi^{(1)} - \phi^{\infty}$ is constant. It is a consequence of (2.10) that the parameters $\lambda^{(k)}$ are bounded, so that it is possible to choose a thinner subsequence such that $\lambda^{(k)} \to \lambda^{\infty}$. On this subsequence $W^{(k)} \to W^{\infty}$ so that functions $q^{\infty}(\delta)$, $l^{\infty}(\delta)$ and $\psi^{\infty}(\delta)$ can be defined analogously to (2.2), (2.3) and (2.4), replacing $f^{(k)}$ by f^{∞} etc. Let $\bar{h}$ satisfy $0 < \bar{h} < \inf(h^{(k)})$, let $\bar{\delta}$ minimize $\psi^{\infty}(\delta)$ on $\|\delta\| \leq \bar{h}$ and let $\bar{x} = x^{\infty} + \bar{\delta}$. For k sufficiently large, it follows that $\|\bar{x} - x^{(k)}\| \leq h^{(k)}$ and so

$$\psi^{(k)}(\bar{x} - x^{(k)}) \geq \psi^{(k)}(\delta^{(k)}) = \phi^{(k)} - \Delta\psi^{(k)}. \tag{3.8}$$

In the limit $k \to \infty$, $f^{(k)} \to f^{\infty}$ etc. and $\bar{x} - x^{(k)} \to \bar{\delta}$, so it follows that $\psi^{(k)}(\bar{x} - x^{(k)}) \to \psi^{\infty}(\bar{\delta})$. Since $\Delta\psi^{(k)} \to 0$ it follows from (3.8) that $\psi^{\infty}(\bar{\delta}) \geq \phi^{\infty} = \psi^{\infty}(0)$. Thus $\delta = 0$

also minimizes $\psi^\infty(\boldsymbol{\delta})$ on $\|\boldsymbol{\delta}\| \leq \bar{h}$, and since the latter constraint is not active it follows that first order conditions hold, which are readily verified to be (3.2) by using (1.4). Note also that in case (ii) it is possible to deduce that second order necessary conditions hold at $\boldsymbol{x}^\infty$.

Of course the existence of a bounded region B which the theorem requires is implied if any level set $\{\boldsymbol{x}: \phi(\boldsymbol{x}) \leq \phi^{(k)}\}$ is bounded. Also the sequence $\{\boldsymbol{x}^{(k)}\}$ is assumed to be infinite; if not, then $\Delta\psi^{(k)} = 0$ for some k, the iteration terminates, and first order conditions are satisfied. One point to emphasize about the theorem is that there are no hidden assumptions such that in the limit certain matrices $\boldsymbol{A}^{(k)}$ have full rank or that multipliers $\boldsymbol{\lambda}^{(k)}$ are bounded. Methods for minimizing max functions and solving nonlinear programming problems can often be proved to be convergent under such assumptions yet can fail in practice. Thus it is important that Theorem 3.1 avoids such assumptions. For instance Han [8] has recently given convergence theorems for min–max optimization when $\boldsymbol{W}^{(k)}$ is replaced by a symmetric matrix $\boldsymbol{B}^{(k)}$. which satisfies $\alpha\boldsymbol{I} \geq \boldsymbol{B}^{(k)} \geq \beta\boldsymbol{I} > 0$. In the most preferable case, $\boldsymbol{B}^{(k)}$ is updated by differences in the gradient of a Lagrangian function and hence depends on Lagrange multiplier estimates. However if these estimates become unbounded, then $\boldsymbol{B}^{(k)}$ is likely to become unbounded so the inequality $\alpha\boldsymbol{I} \geq \boldsymbol{B}^{(k)}$ is not then valid. The other assumption that $\boldsymbol{B}^{(k)} \geq \beta\boldsymbol{I}$ presumably contributes to the global convergence proof as it is related to a Levenberg–Marquardt parameter. However Han does not suggest any way to ensure that these bounds are met. In the method of this paper $\boldsymbol{W}^{(k)}$ can also be replaced by the matrix $\boldsymbol{B}^{(k)}$ and Theorem 3.1 remains valid if $\alpha\boldsymbol{I} \geq \boldsymbol{B}^{(k)}$. In this case the Lagrange multipliers are bounded as a consequence of solving (2.5) and so the possibility of $\boldsymbol{B}^{(k)}$ being unbounded is also less likely. Madsen [9] also gives a global convergence result for L_∞ approximation which is subsumed by Theorem 3.1 in the special case that $\boldsymbol{W}^{(k)} = \boldsymbol{0}$.

4. Rates of convergence

It is well known that a global convergence result for an algorithm is more an assurance that premature failure cannot occur, rather than a guarantee of satisfactory behaviour. A worthwhile aim therefore is to aim to show that an algorithm exhibits a superlinear rate of convergence near the solution. Since the QL method and algorithm (2.9) use second derivative information directly it is important to show that this usually leads to a second order rate of convergence. This can be done and the analysis requires second order optimality conditions for (2.4) or (2.5) to be developed. Also a number of special cases have to be considered, treating both exact penalty function and $\|\cdot\|_1$ applications separately. This would unduly extend this paper so the results are justified in detail in [6]. What can be proved however under mild assumptions, and for most practical problems represented by (1.1), are the following.

(i) The second order sufficient conditions which are assumed to hold are 'almost necessary' in a certain sense and hence likely to hold in practice.

(ii) The QL method based on (2.4) converges at a second order rate if $\boldsymbol{x}^{(k)}$, $\boldsymbol{\lambda}^{(k)}$ is sufficiently close to $\boldsymbol{x}^*$, $\boldsymbol{\lambda}^*$.

(iii) When solving nonlinear programming problems by an exact penalty function, the QL method is equivalent to the SOLVER method applied to the nonlinear programming problem, in a neighbourhood of $\boldsymbol{x}^*$, $\boldsymbol{\lambda}^*$.

However when the general form of algorithm (2.9) which includes the step restriction is considered, two less favourable facts emerge. For smooth optimization ($m = 0$) the result that

$$\phi(\boldsymbol{x}^{(k)} + \epsilon^{(k)}\boldsymbol{s}) = \psi^{(k)}(\epsilon^{(k)}\boldsymbol{s}) + o(\epsilon^{(k)2}) \tag{4.1}$$

can be used to demonstrate that any accumulation point also satisfies second order necessary conditions, which nicely completes the theoretical properties. Unfortunately (4.1) is not valid for NDO so it is not clear whether this conclusion holds for algorithm (2.9) (although it does in case (ii)—see Theorem 3.1). Moreover an observation of Maratos [10] (see [11, 1] is that $\boldsymbol{x}^{(k)}$ can be arbitrarily close to $\boldsymbol{x}^*$ and the unit step $\boldsymbol{x}^{(k)} + \boldsymbol{\delta}^{(k)}$ of the QL method can fail to reduce $\phi(\boldsymbol{x})$. This occurs when there are directions of zero slope at $\boldsymbol{x}^*$ and $\boldsymbol{x}^{(k)}$ is close to the curved surfaces of discontinuity in $\nabla\phi$. It is most likely to arise when $\|\nabla\phi\|$ is large. The *Maratos effect*, as it might be called, may cause the unit step to be rejected and hence obviates the proof of a second order rate of convergence. In practice I have not noticed that this effect retards convergence and usually the unit step reduces $\phi(\boldsymbol{x})$ close to $\boldsymbol{x}^*$, and a second order rate of convergence is observed. My evidence for this is the solution of eight different nonlinear programming test problems using an L_1 exact penalty function [5]. Also Womersley [13] reports NDO calculations with various different second order algorithms on about 30 test problems of various types with the same conclusions. However examples of the Maratos effect have been noticed in practice [1] . Therefore it is preferable to seek some modified form of algorithm (2.9). It is surprising that the Maratos effect has not been noticed previously since NDO problems which are special cases of (1.1) have been studied for some considerable time. It is tempting to ascribe this to the fact that its effect is not often significant in practice. Alternatively one can cite the relative crudity of some earlier algorithms for NDO or the fact that the effect does not arise when there are no feasible directions of zero slope at $\boldsymbol{x}^*$ as in most L_1 and L_∞ best approximation problems in data fitting.

Acknowledgment

It is a pleasure for me to acknowledge the many discussions I have had with G. A. Watson and R. S. Womersley concerning the content of this paper. I am also grateful to the latter for carefully reading the paper and pointing out a number of errors. The paper was prepared whilst I was Visiting Professor at the University of Kentucky, during the academic year 1979–1980. This invitation and the support of NSF Grant ECS-7923272 are gratefully acknowledged.

Appendix

This appendix shows that the QL method which chooses $\boldsymbol{\delta}^{(k)}$ to minimize $\psi^{(k)}(\boldsymbol{\delta})$ in (2.4) is equivalent to the SOLVER method applied to (2.1). If $\boldsymbol{\mu}^{(k)}$ is an estimate of the optimum Lagrange multipliers (obtained by solving (A.1) at iteration $k-1$), then the quadratic program

$$\begin{aligned} &\underset{\boldsymbol{\delta},v}{\text{minimize}} \quad v + \tfrac{1}{2}\boldsymbol{\delta}^{\mathrm{T}}\Big(\sum_i \mu_i^{(k)}(\nabla^2 f^{(k)} + \nabla^2(\boldsymbol{c}^{\mathrm{T}}\boldsymbol{h}_i)^{(k)}\Big)\boldsymbol{\delta}, \\ &\text{subject to} \quad v - (\boldsymbol{g}^{(k)} + \boldsymbol{A}^{(k)}\boldsymbol{h}_i)^{\mathrm{T}}\boldsymbol{\delta} \ge f^{(k)} + \boldsymbol{c}^{(k)\mathrm{T}}\boldsymbol{h}_i + b_i, \quad \forall i \end{aligned} \tag{A.1}$$

determines $\boldsymbol{\delta}^{(k)}$ in the SOLVER method. It is assumed that $\boldsymbol{\lambda}^{(1)} = \boldsymbol{H}\boldsymbol{u}^{(1)}$ and (in an inductive argument) that $\boldsymbol{\lambda}^{(k)} = \boldsymbol{H}\boldsymbol{\mu}^{(k)}$, so that the quadratic term in the objective function of (A.1) can be written as $\frac{1}{2}\boldsymbol{\delta}^{\mathrm{T}}\boldsymbol{W}^{(k)}\boldsymbol{\delta}$ where $\boldsymbol{W}^{(k)}$ is defined as in (2.3). Then writing $w = v + \frac{1}{2}\boldsymbol{\delta}^{\mathrm{T}}\boldsymbol{W}^{(k)}\boldsymbol{\delta}$, (A.1) becomes

$$\begin{aligned} &\underset{\boldsymbol{\delta},w}{\text{minimize}} \quad w, \\ &\text{subject to} \quad w \ge q^{(k)}(\boldsymbol{\delta}) + \boldsymbol{h}_i^{\mathrm{T}}\boldsymbol{l}^{(k)}(\boldsymbol{\delta}) + b_i, \quad \forall i \end{aligned}$$

which is equivalent to

$$\underset{\boldsymbol{\delta}}{\text{minimize}} \quad \max_i q^{(k)}(\boldsymbol{\delta}) + \boldsymbol{h}_i^{\mathrm{T}}\boldsymbol{l}^{(k)}(\boldsymbol{\delta}) + b_i$$

and hence to

$$\underset{\boldsymbol{\delta}}{\text{minimize}}\; q^{(k)}(\boldsymbol{\delta}) + h(\boldsymbol{l}^{(k)}(\boldsymbol{\delta})). \tag{A.2}$$

This is the unrestricted subproblem (2.5). The first order conditions for (A.1) are that

$$\begin{aligned} &\sum_i \mu_i(\boldsymbol{g}^{(k)} + \boldsymbol{A}^{(k)}\boldsymbol{h}_i) = \boldsymbol{0}, \quad \sum_i \mu_i = 1, \\ &\mu_i = 0, \quad i \notin \mathcal{A}^{(k)}, \qquad \boldsymbol{\mu}^{(k)} \le \boldsymbol{0} \end{aligned}$$

which, using $\boldsymbol{\lambda} = \boldsymbol{H}\boldsymbol{\mu}$ and (1.7), are equivalent to

$$\boldsymbol{g}^{(k)} + \boldsymbol{A}^{(k)}\boldsymbol{\lambda} = \boldsymbol{0}, \quad \boldsymbol{\lambda} \in \partial h^{(k)}$$

which are the first order conditions for (A.2). Thus $\boldsymbol{\lambda}^{(k+1)}$ is equivalent to $\boldsymbol{H}\boldsymbol{\mu}^{(k+1)}$ which completes the inductive hypothesis. Thus the equivalence is established. This result generalizes the idea which occurs in nonlinear programming of making a linear approximation $\boldsymbol{l}^{(k)}(\boldsymbol{\delta})$ to constraint functions $\boldsymbol{c}(\boldsymbol{x})$ and including second order terms $\nabla^2 c_i^{(k)}$ in an approximation $q^{(k)}(\boldsymbol{\delta})$ to the objective function $f(\boldsymbol{x})$ in order to give a second order rate of convergence. This idea is shown to be applicable through (1.1) to a wide range of NDO problems.

References

[1] R.M. Chamberlain, C. Lemarechal, H.C. Pedersen and M.J.D. Powell, "The watchdog technique for forcing convergence in algorithms for constrained optimization," Research Report DAMTP 80/NA1 (1980).

[2] F.H. Clarke, "Generalized gradients and applications," *Transactions of the American Mathematical Society* 205 (1975) 247–262.

[3] R. Fletcher, "Methods related to Lagrangian functions," in: P.E. Gill and W. Murray, eds., *Numerical methods for constrained optimization* (Academic Press, London, 1974) pp. 219–239.

[4] R. Fletcher, *Practical methods of optimization*, Vol. I, Unconstrained optimization (Wiley, New York 1980).

[5] R. Fletcher, "Numerical experiments with an L_1 exact penalty function method," in: O.L. Mangasarian, R.R. Meyer and S.M. Robinson, eds., *Nonlinear programming* 4 (Academic Press, New York, 1981) pp. 99–129.

[6] R. Fletcher, *Practical methods of optimization*, Vol. II, Constrained optimization (Wiley, New York, 1981).

[7] R. Fletcher and G. A. Watson, "First and second order conditions for a class of nondifferentiable optimization problems," *Mathematical Programming* 18 (1980) 291–307.

[8] S. P. Han, "Variable metric methods for minimizing a class of nondifferentiable functions," *Mathematical Programming* 20 (1981) 1–13.

[9] K. Madsen, "An algorithm for minimax solution of overdetermined systems of nonlinear equations," *Journal of the Institute of Mathematics and Its Applications* 16 (1975) 321–328.

[10] N. Maratos, "Exact penalty function algorithms for finite dimensional and control optimization problems," Ph.D. Thesis, University of London, London (1978).

] D.Q. Mayne, "On the use of exact penalty functions to determine step length in optimization algorithms," in: G.A. Watson, ed., *Numerical Analysis*, Lecture Notes in Mathematics 773 (Springer, Berlin, 1980) pp. 98–109.

] B.N. Pshenichnyi, "Nonsmooth optimization and nonlinear programming," in: C. Lemarechal and R. Mifflin, eds., *Nonsmooth optimization* (Pergamon, Oxford, 1978) pp. 71–78.

] R.S. Womersley, "Numerical methods for structured problems in non-smooth optimization", Ph.D. Thesis, Department of Mathematics, University of Dundee (1981).

Mathematical Programming Study 17 (1982) 77–90.
North-Holland Publishing Company

A MODIFICATION AND AN EXTENSION OF LEMARECHAL'S ALGORITHM FOR NONSMOOTH MINIMIZATION*

Robert MIFFLIN

Department of Pure and Applied Mathematics, Washington State University, Pullman, WA 99164, U.S.A.

Received 20 November 1980
Revised manuscript received 14 September 1981

An algorithm is given for finding stationary points for constrained minimization problems having locally Lipschitz problem functions that are not necessarily convex or differentiable but are semismooth.

Key words: Nonsmooth Optimization, Nondifferentiable Programming, Constrained Minimization, Semismooth Functions.

1. Introduction

We consider the problem of minimizing f on $S=\{x\in\mathbf{R}^n: h(x)\leq 0\}$ where f and h are real-valued locally Lipschitz continuous functions defined on $\mathbf{R}^n$. We give a modification and an extension of an algorithm due to Lemarechal [2] and show convergence to a stationary point of the problem if f and h also satisfy a weak 'semismoothness' [5, 6] hypothesis that is most likely satisfied by continuous functions arising in practical applications. The method is a feasible point descent method which combines a generalized cutting plane idea with quadratic approximation of some Lagrangian function. Even for the special case of no constraint function h (i.e., $S=\mathbf{R}^n$) and a convex objective function f, as considered in [2], this version differs from the original method, because of its rules for line search termination and the associated updating of the search direction finding subproblem. More specifically, our version does not require a user-specified uniform lower bound on the line search stepsizes. Instead, it uses a 'two-point' line search related to the one introduced in [5]. In this paper, we introduce a general 'α-function' whose values appear both in the subproblem constraints and in the stopping test for the line search. The line search stopping rules provide a direct generalization of those rules found useful for unconstrained minimization of a smooth function. The α-function concept along with the idea of bundling generalized gradients of f from points in S and those of h from points outside S is what deals with nonsmoothness arising either from nondifferentiability of the problem functions or from the presence of a constraint boundary.

*This material is based upon work supported by the National Science Foundation under Grant No. MCS 78–06716.

The algorithm requires a feasible starting point, i.e., an $x_1 \in S$, but requires no knowledge of f at infeasible points as do exact penalty function methods. If no such x_1 is available, then the algorithm can be used to minimize h starting from any point and if h is semiconvex [6] and there exists an $\hat{x}$ such that $h(\hat{x}) < 0$ then, by the convergence theorem in Section 4, the algorithm will find a feasible point in a finite number of iterations.

Also at feasible points the method requires no knowledge of h (other than h being nonpositive) as is required by many feasible point methods. In fact, since $h(x)$ may be replaced by $\max(h(x), 0)$ without changing the feasible set S, the constraint function may be assumed to be zero throughout S. The important question of scaling h (or its components if it is a maximum of several constraint functions) by a positive multiple is not discussed in this paper.

For a locally Lipschitz function F on $\mathbf{R}^n$ let ∂F denote the generalized gradient [1] of F, i.e., for $x \in \mathbf{R}^n$, $\partial F(x)$ is the convex hull (conv) of all limits of sequences of the form $\{\nabla F(x_k)\colon \{x_k\} \to x$ and F is differentiable at each $x_k\}$. Important properties of the point-to-convex set mapping $\partial F(\cdot)$ are uppersemicontinuity and local boundedness. If F is convex ∂F equals the subdifferential, i.e., for each $x \in \mathbf{R}^n$

$$g \in \partial F(x) \quad \text{if and only if} \quad F(y) \geq F(x) + \langle g, y - x \rangle \quad \text{for all } y \in \mathbf{R}^n. \tag{1}$$

If F is continuously differentiable (C^1) ∂F equals the (ordinary) gradient $\{\nabla F\}$. Furthermore, for many other functions F, such as those that are pieced together from C^1 functions, it is possible to determine ∂F or at least to give one element of $\partial F(x)$ at each x. For example, if $F(x) = \max[F_1(x)\ F_2(x), \dots, F_m(x)]$ where each F_i is C^1, then $\partial F(x) = \operatorname{conv}\{\nabla F_i(x)\colon F_i(x) = F(x)\}$. Such examples occur in decomposition, relaxation, duality, and/or exact penalty approaches to solving optimization problems. For further details, generalizations and related results see the references in the comprehensive nonsmooth optimization bibliography in [3] .

We say point $\bar{x} \in S$ is *stationary* for f on S if $0 \in M(\bar{x})$ where

$$M(x) = \begin{cases} \partial f(x), & \text{if } h(x) < 0, \\ \operatorname{conv}(\partial f(x) \cup \partial h(x)), & \text{if } h(x) = 0, \\ \partial h(x), & \text{if } h(x) > 0, \end{cases}$$

because $0 \in M(x^*)$ is a necessary condition for $x^* \in S$ to minimize f on S. However, as shown in [9] using an example of a nonsmooth nonmax function also given in [6], there may exist feasible directions of strict descent at a stationary point. As indicated in [5], it can be shown that the point-to-convex set mapping $M(\cdot)$ inherits the generalized gradient properties of uppersemicontinuity and local boundedness, which are required for showing convergence of the algorithm.

In order to implement the algorithm, we suppose that we have a subroutine that can evaluate a function $g(x) \in M(x)$ for each $x \in \mathbf{R}^n$. For ease of implementation and exposition, we also suppose that $g(x) \in \partial f(x)$ if $h(x)=0$. Of course, we are especially interested in the nonsmooth case where g is discontinuous at stationary points of the constrained minimization problem. Because g may have discontinuities, the algorithm employs a two-point line search. For example, if a line search proceeds along a direction that goes outside S, it must obtain two points, one feasible and the other infeasible, in order to simultaneously maintain feasibility and take the constraint boundary into account properly without requiring knowledge of ∂h at feasible points. Even for line searches along feasible directions, the method may need g-values on both sides of a discontinuity in the gradient of f. For example, if $f(x)=|x|=\max[x, -x]$ for $x \in \mathbf{R}$, then $g(x)=1$ for $x>0$ and $g(x)=-1$ for $x<0$, and we need to know both of these derivatives in order to identify $x=0$ as a minimizing point of f, i.e., in order to conclude that $0 \in \partial f(0)=\operatorname{conv}\{-1, 1\}$. To insure that g-values taken at points near a g-discontinuity are close to being generalized gradients at the discontinuity the algorithm makes use of an α-function that is defined as follows:

Definition. Associated with f, h and g let $\alpha : S \times \mathbf{R}^n \to \mathbf{R}_+$ be a nonnegative-valued function satisfying

$$\{\alpha(x_k, y_k)\} \to 0, \qquad \text{if } \{(x_k, y_k)\} \to (\bar{x}, \bar{x}), \tag{2a}$$

$$\{\alpha(w_k, y_k) - \alpha(x_k, y_k)\} \to 0, \quad \text{if } \{(w_k, x_k, y_k)\} \to (\bar{x}, \bar{x}, \bar{y}), \tag{2b}$$

and

$$\bar{g} \in M(\bar{x}), \text{if } \{(x_k, y_k, g(y_k))\} \to (\bar{x}, \bar{y}, \bar{g}) \quad \text{and} \quad \{\alpha(x_k, y_k)\} \to 0. \tag{2c}$$

$\alpha(x, y)$ is intended to be an indication of how much $g(y) \in M(y)$ deviates from being an element of M at x. However, this measure may be somewhat arbitrary, because a fixed positive multiple of an α-function is also an α-function.

For a convex problem we may take α as a measure of deviation from linearity as indicated in the following lemma. Let S^c be the complement of S and note that S is closed, because h is continuous.

Lemma 1. *Suppose f and h are convex on $\mathbf{R}^n$. Then*

$$\alpha(x, y) = \begin{cases} f(x) - f(y) - \langle g(y), x-y \rangle & \text{for } (x, y) \in S \times S, \\ -h(y) - \langle g(y), x-y \rangle & \text{for } (x, y) \in S \times S^c \end{cases}$$

is nonnegative and satisfies (2).

Proof. If $y \in S$, then $g(y) \in \partial f(y)$ and by the convexity of f, the definition of α and the subgradient inequality (1) we have $\alpha(x, y) \geq 0$. If $y \notin S$, then $g(y) \in \partial h(y)$ and by the convexity of h

$$h(z) \geq h(y) + \langle g(y), z-y \rangle \quad \text{for all } z \in \mathbf{R}^n.$$

Then for $z=x\in S$ we have

$$0\geq h(x)\geq h(y)+\langle g(y), x-y\rangle=-\alpha(x, y).$$

So $\alpha(x, y)$ is nonnegative in either case.

The proofs of properties (2a) to (2c) require separating any sequence $\{y_k\}$ into two subsequences, one in S and the other in S^c, and then making separate arguments for each case as was done above for the nonnegativity proof. We will consider the latter case and omit the former for it has very similar arguments. So, for each sequence $\{y_k\}\to\bar{y}$ considered below, suppose $y_k\notin S$ for all k. Then

$$h(y_k)>0, \qquad g(y_k)\in\partial h(y_k),$$
$$\alpha(x_k, y_k) = -h(y_k)-\langle g(y_k), x_k - y_k\rangle \quad \text{for } x_k \in S, \tag{3}$$

and, by the continuity of h,

$$h(\bar{y})\geq 0. \tag{4}$$

Furthermore, suppose for each sequence $\{x_k\}\to\bar{x}$ considered below that $\{x_k\}\subset S$, so that

$$h(\bar{x})\leq 0. \tag{5}$$

To show (2a) suppose that $\bar{y}=\bar{x}$. Then, by (4) and (5)

$$0\geq h(\bar{x}) = h(\bar{y})\geq 0, \tag{6}$$

The local boundedness of ∂h implies

$$\{-h(y_k)-\langle g(y_k), x_k-y_k\rangle\}\to -h(\bar{y}). \tag{7}$$

Then (2a) follows from (3), (7) and (6).

To show (2b) suppose $\{w_k\}\to\bar{x}$. Since $\alpha(w_k, y_k)-\alpha(x_k, y_k)=\langle g(y_k), x_k-w_k\rangle$, (2b) follows from the local boundedness of ∂h.

To show (2c) suppose $\{\alpha(x_k, y_k)\}\to 0$ and $\{g(y_k)\}\to\bar{g}$. Then

$$\{-h(y_k)-\langle g(y_k), x_k-y_k\rangle\}\to 0,$$

so, by continuity,

$$-h(\bar{y})-\langle\bar{g}, \bar{x}-\bar{y}\rangle=0. \tag{8}$$

Furthermore, by uppersemicontinuity of ∂h, $\bar{g}\in\partial h(\bar{y})$, so, by convexity of h,

$$h(z)\geq h(\bar{y})+\langle\bar{g}, z-\bar{y}\rangle \quad \text{for all } z\in\mathbf{R}^n,$$

which combined with (4) gives

$$h(z)\geq\langle\bar{g}, z-\bar{x}\rangle \quad \text{for all } z\in\mathbf{R}^n. \tag{9}$$

Setting $z=\bar{x}$ gives $h(\bar{x})\geq 0$, which combined with (5) yields

$$h(\bar{x})=0. \tag{10}$$

Finally, (9) and (10) imply

$$h(z) \geq h(\bar{x}) + \langle \bar{g}, z-x \rangle \quad \text{for all } z \in \mathbf{R}^n,$$

which, by convexity of h, *implies that* $\bar{g} \in \partial h(\bar{x})$ and together with (10) implies that $\bar{g} \in M(\bar{x})$ and completes the proof.

For general problems it can be shown that the following function is satisfactory:

$$\alpha(x, y) = \begin{cases} \max[f(x)-f(y)-\langle g(y), x-y \rangle, \beta_0 |x-y|^2] & \text{for } (x, y) \in S \times S, \\ \max[-h(y)-\langle g(y), x-y \rangle, \beta_1 |x-y|^2] & \text{for } (x, y) \in S \times S^c, \end{cases}$$

where β_0 and β_1 are positive parameters and $|\cdot|$ denotes Euclidean norm. Note that if either problem function is known to be convex, then the corresponding β-parameter may be set equal to zero.

To motivate the search direction finding subproblem employed by the algorithm consider the simplest case where f is convex and the problem is unconstrained, i.e., $S = \mathbf{R}^n$. Suppose that f and g have been evaluated at x and y. A polyhedral approximation to $f(x+d)$ which, by convexity, agrees with $f(x)$ and $f(y)$ when $d=0$ and $d=y-x$, respectively, is given by

$$\max[f(x)+\langle g(x), d \rangle, f(y)+\langle g(y), d-y+x \rangle]$$

or, using the definition of α given in Lemma 1,

$$\max[f(x)+\langle g(x), d \rangle, f(x)+\langle g(y), d \rangle - \alpha(x, y)].$$

Minimizing this polyhedral function of d is equivalent to the problem

$$\begin{aligned} &\underset{(w, d) \in \mathbf{R}^{1+n}}{\text{minimize}} && w, \\ &\text{subject to} && w \geq f(x)+\langle g(x), d \rangle, \\ &&& w \geq f(x)+\langle g(y), d \rangle - \alpha(x, y). \end{aligned}$$

A quadratic approximation to $f(x+d)$ is given by

$$f(x)+\langle g(x), d \rangle + \tfrac{1}{2}\langle d, Gd \rangle$$

where G is an n by n matrix approximating the curvature of f, for example, satisfying $G(x-z) = g(x)-g(z)$ where z is some past point (not necessarily y) at which g has been evaluated and there may be some indication that f is smooth along the line segment from z to x. Minimizing the above quadratic function of d is equivalent to the problem

$$\begin{aligned} &\underset{(w, d) \in \mathbf{R}^{1+n}}{\text{minimize}} && w + \tfrac{1}{2}\langle d, Gd \rangle, \\ &\text{subject to} && w \geq f(x)+\langle g(x), d \rangle. \end{aligned}$$

Merging these two problems and letting $v = w - f(x)$ give the following combined polyhedral-quadratic approximation subproblem used by the algorithm:

$$\underset{(v,\, d)\in\mathbf{R}^{1+n}}{\text{minimize}} \quad v+\tfrac{1}{2}\langle d, Gd\rangle+f(x),$$
$$\text{subject to} \quad v\geq\langle g(x), d\rangle,$$
$$v\geq\langle g(y), d\rangle-\alpha(x, y).$$

2. The algorithm

Let m_L and m_R be fixed parameters satisfying $0<m_L<m_R<1$ and (in case the problem is smooth) $m_L<\frac{1}{2}$.

Suppose initially that $x_1\in S$ and let $y_1=x_1$ and G_1 be a positive definite $n\times n$ matrix, such as the identity matrix. Note that, by the definitions of g and M, $g(y_1)\in\partial f(x_1)$ and, by property (2a), $\alpha(x_1, y_1)=0$.

In general, given a positive iteration integer k, a feasible point $x_k\in S$, generalized gradients $g(y_i)\in M(y_i)$ and corresponding scalars $\alpha(x_k, y_i)$ where $y_i\in\mathbf{R}^n$ for $i=1, 2, \ldots, k$, and a positive definite $n\times n$ matrix G_k solve for $(d, v)=(d_k, v_k)\in\mathbf{R}^{n+1}$ the kth quadratic programming subproblem:

$$\text{minimize} \quad \tfrac{1}{2}\langle d, G_k d\rangle+v,$$
$$\text{subject to} \quad -\alpha(x_k, y_i)+\langle g(y_i), d\rangle\leq v \quad \text{for } i=1, 2, \ldots, k.$$

If $v_k=0$ stop. Otherwise, by a line search procedure as given below, find (if possible) two stepsizes t_L and t_R such that $0\leq t_L\leq t_R$ and such that the two corresponding points defined by

$$x_{k+1}=x_k+t_L d_L \quad \text{and} \quad y_{k+1}=x_k+t_R d_k$$

satisfy

$$h(x_{k+1})\leq 0, \tag{11a}$$

$$f(x_{k+1})\leq f(x_k)+m_L t_L v_k, \tag{11b}$$

and

$$-\alpha(x_{k+1}, y_{k+1})+\langle g(y_{k+1}), d_k\rangle\geq m_R v_k. \tag{11c}$$

If the line search procedure is successful, define the $(k+1)$st subproblem by replacing in the subproblem constraints $\alpha(x_k, y_i)$ by $\alpha(x_{k+1}, y_i)$ for $i=1, 2, \ldots, k$, appending the constraint

$$-\alpha(x_{k+1}, y_{k+1})+\langle g(y_{k+1}), d\rangle\leq v$$

and replacing G_k by a positive definite matrix G_{k+1}.

3. Remarks on the algorithm

If (d_k, v_k) solves the kth subproblem, then, necessarily, there exist multipliers λ_{ik} for $i=1, 2, \ldots, k$ such that

$$\lambda_{ik}\geq 0, \tag{12a}$$

$$\sum_{i=1}^{k} \lambda_{ik} = 1, \tag{12b}$$

$$G_k d_k = -\sum_{i=1}^{k} \lambda_{ik} g(y_i) \tag{12c}$$

and

$$v_k = \langle g(y_i), d_k \rangle - \alpha(x_k, y_i) \quad \text{if } \lambda_{ik} > 0. \tag{12d}$$

Combining (12a) to (12d) gives

$$v_k = -\langle G_k d_k, d_k \rangle - \sum_{i=1}^{k} \lambda_{ik} \alpha(x_k, y_i) \tag{12e}$$

or, since G_k is positive semidefinite and λ_{ik} and α are nonnegative

$$v_k = -\langle G_k^{1/2} d_k, G_k^{1/2} d_k \rangle - \sum_{i=1}^{k} \lambda_{ik} \alpha(x_k, y_i) \leq 0 \tag{13}$$

where $G_k^{1/2} G_k^{1/2} = G_k$.

From (12c) and (12e), we conclude that v_k can be thought of as an approximate directional derivative of some Lagrangian function at x_k in the direction d_k. From (13) we conclude that if $v_k \neq 0$ (i.e., the algorithm does not terminate) then the line search procedure is entered with $v_k < 0$. The next result justifies termination when $v_k = 0$.

Lemma 2. *If $v_k = 0$, then x_k is stationary for f on S.*

Proof. If $v_k = 0$, then, from (13) and the nonnegativity of λ_{ik} and α, we have

$$\lambda_{ik} \alpha(x_k, y_i) = 0 \quad \text{for } i = 1, 2, \dots, k \tag{14a}$$

and

$$G_k^{1/2} d_k = 0. \tag{14b}$$

Multiplying (14b) by $G_k^{1/2}$ and combining the result with (12c) gives

$$\sum_{i=1}^{k} \lambda_{ik} g(y_i) = 0. \tag{14c}$$

Furthermore, (14a) implies that $\alpha(x_k, y_i) = 0$ if $\lambda_{ik} > 0$. Thus, by property (2c), $g(y_i)$ is an element of the convex set $M(x_k)$ for each i such that $\lambda_{ik} > 0$. Now, stationary follows from (14c) and the fact that the λ_{ik} form a convex combination.

Note that the above discussion only requires G_k to be positive semidefinite. When G_k is positive semidefinite the subproblem has a bounded solution if there exist d_k and λ_{ik} for $i = 1, 2, \dots, k$ satisfying (12c). Assuming G_k to be positive definite guarantees that (12c) can be satisfied and that the subproblem has a dual, as developed in [2], given by

$$\text{minimize} \quad \frac{1}{2}\left\langle \sum_{i=1}^{k} \lambda_i g(y_i), \sum_{i=1}^{k} \lambda_i G_k^{-1} g(y_i) \right\rangle + \sum_{i=1}^{k} \lambda_i \alpha(x_k, y_i),$$

$$\text{subject to} \quad \sum_{i=1}^{K} \lambda_i = 1 \quad \text{and} \quad \lambda_i \geq 0 \quad \text{for } i = 1, 2, \ldots, k.$$

Elsewhere, we will report on how to extend the numerically stable constrained least squares algorithm in [7] for solving more general quadratic programming problems. A specialization of this new method will result in a reliable method for solving subproblem dual given here.

The constraints in the subproblem are present to determine the active generalized gradients (and their multipliers) at optimiality and the objective matrix G_k is present to represent the curvature (if any) of a Lagrangian associated with f and h with respect to directions that are orthogonal to all of the active gradients. Current research centers on finding out when and how

(a) to make some type of variable metric update of G_k based on differencing nonnegative combinations of gradients, i.e., differencing approximate gradients of a Lagrangian;

(b) to aggregate, reduce, or restart the constraint 'bundle'.

We speculate that the above combination of cutting-plane and quasi-Newton aspects are needed for the algorithm to have the potential for some kind of superlinear convergence. The details for bounding the number of subproblem constraints and for updating G_k in order to obtain superlinear convergence in the one-dimensional unconstrained convex case where $n=1$ and f is convex on $S=\mathbf{R}$ are given in [4]. This research (incomplete in the higher dimensional and constrained cases) is not discussed here, for the purpose of this paper is to establish some general convergence theory.

Line search stopping criteria (11a) and (11b) guarantee that x_{k+1} is feasible with a sufficiently smaller objective value than x_k and (11c) makes (d_k, v_k) sufficiently infeasible in subproblem $k+1$, because $v_k<0$ and $m_R<1$ imply that

$$-\alpha(x_{k+1}, y_{k+1}) + \langle g(y_{k+1}), d_k \rangle \geq m_R v_k > v_k. \tag{15}$$

Criteria (11b) and (11c) are generalizations of the line search stopping rules found useful in smooth minimization. (See [8], for example.) The smooth case appears when $g(x_k+td_k)=\nabla f(x_k+td_k)$ for all t, $v_k=\langle \nabla f(x_k), d_k\rangle$ (the exact directional derivative of f at x_k in direction d_k) and $t_R=t_L>0$, for then $y_{k+1}=x_{k+1}\neq x_k$ and, by (2a), $\alpha(x_{k+1}, y_{k+1})=0$. In general, discontinuities in g are detected and dealt with by allowing $y_{k+1}\neq x_{k+1}$, but imposing an upper bound on $\alpha(x_{k+1}, y_{k+1})$ via (15) which converges to zero if d_k and v_k converge to zero.

If f and h are weakly upper semismooth on $\mathbf{R}^n$ (see Appendix for the definition), then the following line search procedure either finds t_L and t_R or generates an increasing sequence $\{t_L\}$ such that $\{x_L+t_L d_k\}\subset S$ and $\{f(x_k+t_L d_k)\}\to -\infty$.

Line Search Procedure

Let ρ be a fixed parameter satisfying $0<\rho<\frac{1}{2}$.

Set $t=0$, $t_L=0$, and $t_u=+\infty$.

Loop: If $h(x_k+td_k)\leq 0$ and $f(x_k+td_k)\leq f(x_k)+m_L t v_k$
set $t_L=t$. Otherwise set $t_u=t$.
If $-\alpha(x_k+t_L d_k, x_k+td_k)+\langle g(x_k+td_k), d_k\rangle\geq m_R v_k$
set $t_R=t$ and return. Otherwise
if $t_u=+\infty$, choose $t\geq t_L+\rho$ by some extrapolation procedure,
or if t_u is finite choose $t\in[t_L+\rho(t_u-t_L), t_u-\rho(t_u-t_L)]$
by some interpolation procedure.
Go to Loop.

If the return never occurs, then the successive choices of t either cause $t_L\uparrow+\infty$ if t_u remains $+\infty$ or cause $(t_u-t_L)\downarrow 0$ if t_u ever becomes finite. In the former case, f is unbounded on the feasible set, because $m_L>0$ and $v_k<0$ imply that $m_L t_L v_k\downarrow-\infty$. In the latter case, a contradiction can be established as in [5] if f and h are weakly upper-semismooth. Hence, if the return does not occur after a finite number of loops, then the unbounded case obtains when the problem functions satisfy this additional semismoothness hypothesis. This result depends on the α-function property (2a) and the parameter inequality $m_L<m_R$ as well as the line search procedure rules.

The possibility where $\langle g(x_k), d_k\rangle\geq m_R v_k$ requires special consideration. This case occurs if and only if the loop is executed only once and $t_R=t_L=0$. Then $y_{k+1}=x_{k+1}=x_k$ and, by property (2a), the appended subproblem constraint is $\langle g(x_{k+1}), d>\leq v$. So, if at the next iteration $v_{k+1}<0$, then, since $m_R<1$, we have

$$\langle g(x_{k+1}), d_{k+1}\rangle\leq v_{k+1}<m_R v_{k+1}.$$

Thus, defining $K_0=\{l: \langle g(x_l), d_l\rangle<m_R v_l\}$ we have that if $k\notin K_0$, then $x_{k+1}=x_k$ and either $v_{k+1}=0$ or $k+1\in K_0$. Therefore, if the algorithm does not terminate K_0 is an infinite set.

The final result of this section shows that in the convex case if the first positive t-value chosen does not exceed 1 and if it fails to become t_L, then the line search stops after only one evaluation of (h, f, g) with t_R equal to this t-value. On the other hand, if this first positive t-value does become t_L, then this value is a positive lower bound on the terminal value of t_L.

Lemma 3. *Suppose f and h are convex on $\mathbf{R}^n$ and the form of α is that given in Lemma 1. Also, suppose that if the line search does not terminate with $t_R=t_L=0$, then t is set equal to $t_1\in(0, 1]$ for the next loop.*

(a) *If t_u is set equal to t_1 at the next loop, then termination occurs with $t_R=t_1$ and $t_L=0$.*

(b) *Otherwise (i.e., t_L is set equal to t_1 at the next loop), then at termination $t_1\leq t_L\leq t_R$.*

Proof. (a) If the next loop execution leaves $t_L=0$ and sets $t_u=t_1$, then either

$$h(x_k+t_1d_k)>0 \tag{16a}$$

or

$$f(x_k+t_1d_k)>f(x_k)+m_L t_1 v_k.$$

We give the proof in the former case, for the latter case is similar and essentially contained in [2; p. 22]. So in the former case from the definition of α

$$-\alpha(x_k+t_Ld_k, x_k+t_1d_k)=h(x_k+t_1d_k)-t_1\langle g(x_k+t_1d_k), d_k\rangle$$

which combined with (16a) implies

$$-\alpha(x_k+t_Ld_k, x_k+t_1d_k)+\langle g(x_k+t_1d_k), d_k\rangle>(1-t_1)\langle g(x_k+t_1d_k), d_k\rangle. \tag{16b}$$

Now, by the convexity of h,

$$h(x_k+t_1d_k)-t_1\langle g(x_k+t_1d_k), d_k\rangle\le h(x_k),$$

and, since $t_1>0$, $h(x_k)\le 0$, and $h(x_k+t_1d_k)>0$, we have

$$0<[h(x_k+t_1d_k)-h(x_k)]/t_1\le\langle g(x_k+t_1d_k), d_k\rangle. \tag{16c}$$

The assumption that $t_1\le 1$ combined with (16b) and (16c) gives

$$-\alpha(x_k+t_Ld_k, x_k+t_1d_k)+\langle g(x_k+t_1d_k), d_k\rangle>0$$

which implies termination with $t_R=t_1$, because $m_R v_k<0$.

(b) The conclusion follows from the fact that the successive values for t_L are increasing.

4. Convergence

Suppose that each execution of the line search procedure is finite and that the algorithm does not terminate. Then $v_k<0$ for all k and K_0 is an infinite set. The three lemmas that we prove in this section establish part (a) of the following convergence theorem. Part (b) follows from part (a), because, if f and h are semiconvex (see Appendix for the definition) and a constraint qualification is satisfied, then any stationary point is a minimizing point [6] and because every accumulation point of $\{x_k\}\subset S$ has the same f-value due to the monotonicity of $\{f(x_k)\}$.

Theorem 1. *Suppose $\{x_k\}$, $\{y_k\}$ and $\{G_k\}$ are bounded with $\{G_k\}$ uniformly positive definite. Then*

(a) *at least one of the accumulation points of $\{x_k\}$ is stationary for f on S; and*

(b) *if f and h are semiconvex on $\mathbf{R}^n$ and there exists an $\hat{x}\in\mathbf{R}^n$ such that $h(\hat{x})<0$, then every accumulation point of $\{x_k\}$ minimizes f on S.*

Remark. If $\{x\in\mathbf{R}^n: x\in S,\ f(x)\le f(x_1)\}$ is bounded, then $\{x_k\}$ is bounded and $\{y_k\}$ can be made bounded by choosing an additional parameter $\epsilon>0$ and imposing the additional line search requirement that

$$|y_{k+1}-x_{k+1}|=(t_R-t_L)|d_k|\le\epsilon.$$

For weakly uppersemismooth functions, it is possible to satisfy this condition and (11) simultaneously after a finite number of line search loops.

Let $\gamma_k>0$ be the smallest eigenvalue of the positive definite matrix G_k. Consider the following assumption that is trivially satisfied if the matrix sequence $\{G_k\}$ is uniformly positive definite:

If $\{d_k\}_{k\in K_0}$ has no zero accumulation point, then

$$\{\gamma_k\}_{k\in K_0} \text{ has no zero accumulation point.} \tag{17}$$

Lemma 4. *Suppose* (17) *holds and* $\{x_k\}$ *and* $\{y_k\}$ *are uniformly bounded. Then* $\{d_k\}_{k\in K_0}$ *has at least one zero accumulation point.*

Proof. Suppose for purposes of a proof by contradiction that there exists a positive number δ such that

$$|d_k|\ge\delta>0 \quad \text{for all } k\in K_0. \tag{18}$$

Then, by (17), there exists a positive number γ such that

$$\gamma_k\ge\gamma>0 \quad \text{for all } k\in K_0. \tag{19}$$

Combining (13), the nonnegativity of λ_{ik} and α, (19) and (18) gives

$$v_k\le-\gamma|d_k|^2\le-\gamma\delta|d_k|\le-\gamma\delta^2<0 \quad \text{for all } k\in K_0. \tag{20}$$

By the Cauchy–Schwarz inequality and the definition of K_0

$$-|g(x_k)||d_k|\le\langle g(x_k), d_k\rangle< m_R v_k \quad \text{for all } k\in K_0.$$

which combined with the left-most inequality of (20) gives

$$|d_k|<|g(x_k)|/(m_R\gamma) \quad \text{for all } k\in K_0.$$

Thus, since $\{x_k\}$ is assumed bounded and $M(\cdot)$ is locally bounded, $\{g(x_k)\}$ and, hence $\{d_k\}_{k\in K_0}$ and $\{v_k\}_{k\in K_0}$ are bounded. Let $\bar{v}$ and $\bar{d}$ be accumulation points of $\{v_k\}_{k\in K_0}$ and $\{d_k\}_{k\in K_0}$, respectively. Then, by (20),

$$\bar{v}\le-\gamma\delta^2<0. \tag{21}$$

By (11b) and by (20) for $k\in K_0$

$$f(x_{k+1})-f(x_k)\le m_L t_L v_k\le-m_L t_L\gamma\delta|d_k|$$

or, since $x_{k+1}=x_k+t_L d_k$,

$$f(x_{k+1})-f(x_k)\le-m_L\gamma\delta|x_{k+1}-x_k|. \tag{22}$$

Note that (22) also holds for $k \notin K_0$, because in this case $x_{k+1} = x_k$.

For any $p > k+1$, (22) and the triangle inequality imply

$$f(x_p) - f(x_{k+1}) = \sum_{j=k+1}^{p-1} f(x_{j+1}) - f(x_j) \leq -m_L \gamma \delta \sum_{j=k+1}^{p-1} |x_{j+1} - x_j|$$
$$\leq -m_L \gamma \delta |x_p - x_{k+1}|.$$

As f is continuous and $\{x_k\}$ is assumed bounded, the monotone nonincreasing sequence $\{f(x_k)\}$ is bounded from below and, hence, there exists an $\bar{x} \in S$ such that

$$\{x_k\} \to \bar{x}. \tag{23}$$

Also, for any $p \geq k+1$ we have, by the pth subproblem feasibility, that

$$-\alpha(x_p, y_{k+1}) + \langle g(y_{k+1}), d_p \rangle \geq v_p. \tag{24}$$

Subtracting (11c) from (24) gives

$$\alpha(x_{k+1}, y_{k+1}) - \alpha(x_p, y_{k+1}) + \langle g(y_{k+1}, d_p - d_k \rangle \leq v_p - m_R v_k. \tag{25}$$

Now choose p and k in K_1, an infinite subset of K_0, where $\{y_{l+1}\}_{l \in K_1} \to \bar{y}$ $\{d_l\}_{l \in K_1} \to \bar{d}$ and $\{v_l\}_{l \in K_1} \to \bar{v}$, so that from (23), (25), property (2b) and the boundedness of $\{g(y_{k+1})\}$ we have

$$0 \leq \bar{v} - m_R \bar{v} = (1 - m_R)\bar{v}.$$

Since $m_R < 1$, this implies that $\bar{v} \geq 0$, which contradicts (21) and completes the proof.

Lemma 5. *Suppose that* $K_1 \subseteq K_0$ *is such that* $\{d_k\}_{k \in K_1} \to 0$, $\{G_k d_k\}_{k \in K_1} \to 0$ *and* $\{\langle g(x_k), d_k \rangle\}_{k \in K_1} \to 0$. *Then* $\{v_k\}_{k \in K_1} \to 0$ *and*

$$\left\{ \sum_{i=1}^{k} \lambda_{ik} \begin{pmatrix} \alpha(x_k, y_i) \\ g(y_i) \end{pmatrix} \right\}_{k \in K_1} \to \begin{pmatrix} 0 \\ 0 \end{pmatrix}$$

where $\lambda_{ik} \geq 0$ *for* $i = 1, 2, \ldots, k$ *and* $\sum_{i=1}^{k} \lambda_{ik} = 1$.

Proof. The conclusions follow from the hypotheses, the definition of K_0, (12) and (13).

Lemma 6. *In addition to the hypotheses of Lemma 5, suppose that* $\{x_k\}_{k \in K_1}$ *and* $\{y_k\}_{k \in K_1}$ *are bounded and let* $\bar{x}$ *be any accumulation point of* $\{x_k\}_{k \in K_1}$. *Then* $\bar{x}$ *is stationary for* f *on* S.

Proof. As in the proof of Theorem 5.2 in [5], depending on the local boundedness and uppersemicontinuity of M and the fact that a convex combination of vectors in $\mathbf{R}^{n+1}$ can be expressed as a convex combination of $n+2$ or fewer of the vectors, Lemma 5 implies the existence of a positive integer $m \leq n+2$, an infinite subset $K_2 \subset K_1$ and convergent subsequences

$$\{(y_k^i, g(y_k^i))\}_{k\in K_2} \to (y^i, g^i) \in \mathbf{R}^n \times M(y^i) \quad \text{and} \quad \{\mu_k^i\}_{k\in K_2} \to \mu^i \geq 0$$

for $i = 1, \ldots, m$ such that

$$\sum_{i=1}^{m} \mu^i = 1, \qquad \sum_{i=1}^{m} \mu^i g^i = 0,$$

and for $i = 1, 2, \ldots, m$ $\{\alpha(x_k, y_k^i)\}_{k\in K_2} \to 0$ if $\mu^i > 0$. Now, because $\{x_k\}_{k\in K_2} \to \bar{x} \in S$, stationarity of $\bar{x}$ follows from property (2c) as in the proof of Lemma 2.

5. A generalization of the α-function class

Consider $\alpha(x, y)$ defined in Lemma 1 and note that if $y \in S$ and $f(x) \leq f(y)$, then

$$\alpha(x, y) = f(x) - f(y) - \langle g(y), x - y\rangle \leq |g(y)||x - y|$$

or if $y \notin S$, then

$$\alpha(x, y) = -h(y) - \langle g(y), x - y\rangle < |g(y)||x - y|,$$

so we may want the general class of α-functions to include $\beta|g(y)||x-y|$ where β is a positive parameter. Although such a function satisfies (2a) and (2b), it does not satisfy (2c), since when $\bar{x} \neq \bar{y}$ we cannot conclude that $\bar{g} \in M(\bar{x})$ if $\{\beta|g(y_k)||x_k - y_k|\} \to 0$, because $\{g(y_k)\} \to \bar{g} = 0$. However, when $\bar{g} = 0$, $\bar{y}$ is stationary if $\bar{y} \in S$. To force feasibility of $\bar{y}$, we could consider functions such as $\max[\beta_1|g(y)||x - y|, \beta_2 h(y)]$ where β_1 and β_2 are positive parameters. Then, in order to include such functions in the α-class, weaken (2c) as follows:

$$\text{either } \bar{g} \in M(\bar{x}) \text{ or } \bar{y} \text{ is stationary}$$
$$\text{if } \{(x_k, y_k, g(y_k))\} \to (\bar{x}, \bar{y}, \bar{g}) \quad \text{and} \quad \{\alpha(x_k, y_k)\} \to 0. \qquad (2c)'$$

If the definition of the α-class is so generalized, then it is easy to show that the following weakened version of Lemma 2 holds.

Lemma 2′. *If $v_k = 0$, then either x_k or y_i for some $i \in \{1, 2, \ldots, k\}$ is stationary for f on S.*

Furthermore, Lemma 6 may be modified in a similar manner so that the following convergence result holds.

Theorem 1′. *Suppose $\{x_k\}$, $\{y_k\}$ and $\{G_k\}$ are bounded with $\{G_k\}$ uniformly positive definite. Then either $\{x_k\}$ or $\{y_k\}$ has at least one accumulation point that is stationary for f on S.*

Appendix

A function $F : \mathbf{R}^n \to \mathbf{R}$ is *weakly uppersemismooth* [5] *at* $x \in \mathbf{R}^n$ if

(a) F is Lipschitz continuous on a ball about x;

(b) for each $d \in \mathbf{R}^n$ and for any sequences $\{t_k\} \subset \mathbf{R}_+$ and $\{g_k\} \subset \mathbf{R}^n$ such that $\{t_k\} \downarrow 0$ and $g_k \in \partial F(x + t_k d)$ it follows that

$$\liminf_{k\to\infty} \langle g_k, d\rangle \geqq \limsup_{t \downarrow 0} [F(x+td)-F(x)]/t.$$

It can be shown that the right hand side of the above inequality is in fact equal to

$$F'(x;d) = \lim_{t \downarrow 0} [F(x+td) - F(x)]/t,$$

the *directional derivative* of F at x in the direction d.

The class of weakly uppersemismooth functions strictly contains the class of semismooth [6] functions. This latter class is closed under composition and contains convex, concave, C^1 and many other locally Lipschitz functions such as ones that result from piecing together C^1 functions as in [9].

A function $F : \mathbf{R}^n \to \mathbf{R}$ is *semiconvex* [6] at $x \in \mathbf{R}^n$ if

(a) F is Lipschitz continuous on a ball about x and for each $d \in \mathbf{R}^n$, $F'(x; d)$ exists and satisfies

(b) $F'(x; d) = \max[\langle g, d\rangle : g \in \partial F(x)]$,

(c) $F'(x; d) \geq 0$ implies $F(x+d) \geq F(x)$.

An example of a nondifferentiable nonconvex function that is both semismooth and semiconvex is $\log(1+|x|)$ for $x \in \mathbf{R}^n$.

References

[1] F.H. Clarke, "Generalized gradients and applications", *Transactions of the American Mathematical Society* 205 (1975) 247–262.

[2] C. Lemarechal, "Nonsmooth optimization and descent methods", RR-78-4, International Institute for Applied Systems Analysis, Laxenburg, Austria (1978).

[3] C. Lemarechal and R. Mifflin, eds., *Nonsmooth optimization* (Pergamon Press, Oxford, 1978).

[4] C. Lemarechal and R. Mifflin, "Global and superlinear convergence of an algorithm for one-dimensional minimization of convex functions", TR-81-3, Department of Pure and Applied Mathematics, Washington State University, Pullman, WA (1981).

[5] R. Mifflin, "An algorithm for constrained optimization with semismooth functions", *Mathematics of Operations Research* 2 (1977) 191–207.

[6] R. Mifflin, "Semismooth and semiconvex functions in constrained optimization", *SIAM Journal of Control and Optimization* 15 (1977) 959–972.

[7] R. Mifflin, "A stable method for solving certain constrained least squares problems", *Mathematical Programming* 16 (1979) 141–158.

[8] M.J.D. Powell, "Some global convergence properties of a variable metric algorithm for minimization without exact line searches", In: R. Cottle and C.E. Lemke, eds., *Nonlinear programming* (American Mathematical Society, Providence, RI, 1976) pp. 53–72.

[9] R.S. Womersley, "Optimality conditions for piecewise smooth functions", *Mathematical Programming Study* 17 (1982) 13–27 [This Volume.].

Mathematical Programming Study 17 (1982) 91–102.
North-Holland Publishing Company

A PROBLEM IN CAPILLARITY AND PLASTICITY

Gilbert STRANG*

Massachusetts Institute of Technology, Cambridge, MA, U.S.A.

Roger TEMAM

Université de Paris-Sud, Paris, France

Received 4 September 1980

In two different applications we meet the problem of minimizing $\iint|\nabla u|\,dx\,dy$ subject to a linear constraint on u. We establish that the solutions come from characteristic functions, and we compute the minimum value—which represents the critical contact angle in capillarity and the collapse factor in plasticity. The analysis is a continuous analogue of the max flow-min cut theorem, and a future publication of the first author will extend that theorem to flows through a domain, depending as here on the coarea formula.

Key words: Convex Analysis, Infinite-dimensional Programming, Bounded Variation, Capillarity, Plasticity.

1. Introduction

This note is about two problems that turn out to be the same. It is a coincidence but not an accident. They both involve a transition from equilibrium to collapse, manifested in one case by a liquid that climbs the sides of a pipe and in the other by a solid that falls apart. In each case there is a scalar multiplying the inhomogeneous terms in the equation and boundary condition, and the object is to find the critical value of that parameter—below the critical value the problem has a solution, and above it the solution breaks down.

The next section describes the problem of capillarity, in which the parameter is the angle γ (governed by surface tension) between the liquid and the walls. The surface of the liquid obeys the Young–Laplace equation; its mean curvature is constant. In the absence of gravity this constant is a multiple of $\cos\gamma$, and as the curvature increases the surface begins climbing the walls. At some point (which can be reproduced by an experiment in space) the surface breaks down completely and the equation has no solution. Concus and Finn [2, 3] were the first to analyze this possibility, and later Giusti [8, 9] found a formula for the critical value of $\cos\gamma$; we want to repeat the essential steps in his argument, linking it to an extremal problem in the space of functions of bounded variation.

In the example from plasticity the underlying geometry is the same—an infinitely long pipe with cross-section Ω. It is composed of a material that behaves elastically provided the stresses are below unity; at that point it

* The first author gratefully acknowledges the support of the National Science Foundation (MCS 78-12363) and the Army Research Office (DAAG29-K0033).

becomes plastic and cannot accept more stress. As the load increases and the material tries to maintain equilibrium, more and more of the pipe reaches this yield state. Ultimately there are no admissible stresses that equilibrate the load, and collapse is the only alternative. For loads that act parallel to the axis this problem is called anti-plane shear, and one example was studied in [11]. (It led to the classical isoperimetric problem, minimizing perimeter/area, but subject to constraints.) Here we push in one direction on the surface of the pipe and in the other direction on the interior, balancing the two so that the pipe doesn't move—but the shear stresses increase until they are unacceptable. Mathematically, a certain feasible set becomes empty, and it is the dual which coincides with the variational problem that arises in capillarity:

$$\text{Minimize} \quad \iint_{\Omega} |\text{grad } u|,$$

$$\text{subject to} \quad \int_{\Gamma} u - \frac{|\Gamma|}{|\Omega|} \iint_{\Omega} u = 1.$$

The minimum gives the critical value of $\cos \gamma$ in one problem and of the surface traction in the other.

The computation of this minimum is made possible by the following fact: *The extremal u is the characteristic function of some subset $E \subset \Omega$.* Or more precisely, since the minimum may not be attained, there is a minimizing sequence of multiples of characteristic functions. (In case it is not attained we have a collapse load without a 'collapse mechanism', a combination which has been conjectured in the theory of plasticity.) It is the co-area formula of Fleming and Rishel [6, 7] and Federer [4] which establishes that the characteristic functions are extremal, and we contribute one small extension to that result: These functions remain extremal if the definition of $|\text{grad } u|$ is changed from $(u_x^2 + u_y^2)^{1/2}$ to any other norm in R^2. In applications to plasticity, that means that the yield condition need not be isotropic—it could be $\max |\sigma_{ij}| \leq 1$—and the yield surface need not be a sphere.

We are grateful to Herbert Federer and Wendell Fleming, especially for discussing and confirming this extension, and to Robert Brown for explaining the problem of capillarity and the shape of a meniscus.

2. Capillarity without gravity

We are given a hollow tube with axis in the z-direction and with cross-section Ω in the x–y plane. The boundary of Ω is Γ. The tube contains a liquid whose surface height is the unknown $u(x, y)$, and the physics of the interface between liquid, walls, and 'air' fixes the contact angle γ along Γ. It is measured between the normal to the surface and the normal to the wall, so that $\gamma = \pi/2$ implies a flat surface without capillary effects; we suppose that $\gamma \leq \pi/2$. Thus

$$\frac{(u_x, u_y, 1)\cdot(n_x, n_y, 0)}{(1+u_x^2+u_y^2)^{1/2}} = \cos\gamma.$$

Writing W for the denominator $(1+|\nabla u|^2)^{1/2}$, this becomes

$$\frac{\operatorname{grad} u \cdot n}{W} = \cos\gamma \quad \text{for } x, y \text{ on } \Gamma. \tag{1}$$

In the interior, the surface satisfies the Young–Laplace equation

$$\operatorname{div}\left(\frac{\operatorname{grad} u}{W}\right) = 2H. \tag{2}$$

We recognize in (1)–(2) the Euler equation and natural boundary condition for the variational problem:

$$\text{Minimize} \iint_\Omega W\, dx dy - \cos\gamma \int_\Gamma u\, ds, \quad \text{with} \iint_\Omega u \text{ given.}$$

The constant mean curvature H becomes the Lagrange multiplier for this constraint, and it is determined from γ by the divergence theorem:

$$\iint_\Omega 2H\, dx dy = \int_\Gamma \cos\gamma\, ds, \quad \text{or} \quad 2H = \frac{|\Gamma|}{|\Omega|}\cos\gamma.$$

Therefore, with this H, the capillarity problem becomes:

$$\text{Minimize} \iint_\Omega W\, dx dy - \cos\gamma\, L(u),$$

$$\text{with } L(u) = \int_\Gamma u\, ds - \frac{|\Gamma|}{|\Omega|}\iint_\Omega u\, dx dy.$$

This is close to Plateau's minimal surface problem, differing only in the presence of the linear term L and the absence of prescribed boundary conditions on u. For $\cos\gamma = 0$, the minimum is attained as we expect by the flat surface $u =$ constant. (Note that L vanishes for this u). In all cases u is determined only up to a constant, since W and L are unchanged if u is replaced by $u + u_0$—and u_0 is determined so as to give the integral of u its prescribed value.

Our goal is to decide whether the minimization problem is bounded or unbounded, that is, *whether or not the minimum is* $-\infty$. This will be the test for collapse, and it depends on γ. For $\cos\gamma$ below a critical value (which is determined by the shape of Ω) the minimum is finite and Giusti and Finn have constructed a solution—a function u at which the minimum is attained, and (1) and (2) are satisfied. For $\cos\gamma$ above this critical value, the term $L(u)$ brings the minimum to $-\infty$ and the equations have no solution. (For $\gamma = \gamma_{\text{critical}}$ we refer to [9].) Thus we are studying only a question of 'yes' or 'no', boundedness or unboundedness, and this allows a simplification of the integrand: The function W can be replaced by $|\nabla u|$. For this we note that

$$|\nabla u| < (1+|\nabla u|^2)^{1/2} = W \leq 1 + |\nabla u|.$$

Therefore the integrals of W and $|\nabla u|$ differ by less than a fixed constant, namely the area of Ω:

$$\iint |\nabla u| < \iint W \leq \iint (1+|\nabla u|) = |\Omega| + \iint |\nabla u|.$$

It follows that the minimum remains finite or remains infinite when we move to the more homogeneous problem,

$$\text{Minimize} \iint_{\Omega} |\nabla u| - \cos\gamma\, L(u).$$

The homogeneity is straightforward; if u is amplified by $\alpha > 0$, then so is the value of the functional to be minimized. Therefore if the functional is negative for any u, its minimum is $-\infty$. In other words, the borderline falls at the point

$$\cos\gamma_{\text{critical}} = \inf_{L(u)=1} \iint |\nabla u|. \tag{3}$$

For $\cos\gamma$ below this infimum we have $\cos\gamma L(u) \leq \iint |\nabla u|$, and above it the sign is reversed for some u; for that u the functional is negative, and then the minimum is $-\infty$.

Our problem is now to compute the infimum in (3). The proper space of admissible functions, for this problem and also for the original minimization (as well as for the classical minimal surfaces), is the space BV of functions of bounded variation. This contains all functions u in L^1 whose gradient ∇u is a bounded measure. In other words, it includes the Sobolev space $W^{1,1}$ (in which the derivatives u_x and u_y are integrable functions), and it also includes some discontinuous u—most importantly, the characteristic functions $u = \chi_E$ of reasonable sets E in Ω. In this case $u = 1$ in E and $u = 0$ outside E, so that $\nabla u = 0$ except on the boundary ∂E; along the boundary we have a 'line of δ-functions', in analogy with the derivative of a step function in one variable. In the latter case, a unit step at x_0 gives

$$\int (du/dx) = \int \delta(x - x_0) = 1,$$

the standard value for the total variation of u. In our two-dimensional case there is still an integral along the boundary ∂E, or at least along that part which lies in Ω:

$$\iint |\nabla u| \, dx dy = \int_{\partial E \cap \Omega} ds$$

$$= \text{relative perimeter of } E = |\partial E \setminus \Gamma|. \tag{4}$$

This is the total variation of the characteristic function $u = \chi_E$, and u belongs to BV provided the perimeter of E is finite. Such an E is called a Caccioppoli set, and the total variation norm of the function is $\|\chi_E\|_{BV} = |\partial E \setminus \Gamma|$.

Roughly speaking, BV is the largest class of functions for which $\iint |\nabla u| < \infty$. By a trace theorem of Miranda, the boundary values of u lie in $L^1(\Gamma)$ so that the linear part $L(u)$ is also finite and our functional is well defined. (Here we take Γ to be piecewise C^1.) This definition of BV was given by Tonelli and Cesari, and in fact it would be more proper to reverse (4) and compute the perimeter of E (relative to Ω) as the total variation of $u = \chi_E$. This total variation, given for smooth functions by $\iint |\nabla u|$, is defined for any integrable u by

$$\|u\| = \sup\left\{\iint u \sum \partial g_i/\partial x_i : g_i \in C_0^1(\Omega), \sum g_i^2 \le 1\right\}.$$

If this is finite, then u is in BV. In fact nothing is changed by adding a constant to u, and $\|u\|$ is the norm in the quotient space BV/R. It is over this space, which neglects constant functions, that we minimize $\|u\|$ subject to $L(u) = 1$.

Example. Suppose Ω is the unit square. Then one possible candidate in the minimization is $u = \chi_E$, where E is a small triangle in the corner—say with sides of length ϵ along the x and y axes. Since $u = 1$ in the triangle and $u = 0$ elsewhere,

$$\int_\Gamma u \, ds = 2\epsilon, \quad \iint_\Omega u = \text{area of } E = \frac{\epsilon^2}{2},$$

$$\iint |\nabla u| = \text{hypotenuse of } E = \sqrt{2}\,\epsilon.$$

For this function we find

$$\frac{\iint |\nabla u|}{L(u)} = \frac{\sqrt{2}\,\epsilon}{2\epsilon - 2\epsilon^2}.$$

As $\epsilon \to 0$ the ratio approaches $1/\sqrt{2}$, and the critical cosine cannot be larger than this; it might be smaller if this special choice of u, concentrated in the corner, could be improved. However Concus and Finn observed that for any angle $\gamma \ge 45°$ a spherical surface will satisfy both the equation and boundary condition. Therefore 45° is the critical angle for a square.

Remark 1. This is an instance in which our reduced minimization, with ∇u replacing W, has no solution at the critical angle 45°; the minimum is finite but it is not attained. (The minimizing sequence of characteristic functions disappears into the corner as $\epsilon \to 0$.) In capillarity the fluid climbs infinitely high only at the four corners, and in plasticity there is no collapse mechanism u at the limit load. On the other hand, the original problem with the extra 1 which appears in W has no difficulty; for it the minimum is attained.

Remark 2. We could have rescaled u so as to achieve $L(u) = 1$, and then computed $\iint |\text{grad } u|$ as required in (3). Instead it was more convenient simply to calculate the ratio $\iint |\text{grad } u|/L(u)$, recognizing that the ratio is unchanged when u is replaced by αu, $\alpha > 0$. In other words, it is immediate that the infimum in (3) is identical to

$$\cos \gamma_{\text{critical}} = \inf_{L(u)>0} \frac{\iint |\text{grad } u|}{L(u)}. \tag{5}$$

Now we turn to the essential point. *The characteristic functions* $u = \chi_E$ *are not only convenient candidates in the minimization* (5), *they are the only candidates that need to be considered.* As in the example, there is always a minimizing sequence of characteristic functions; and if the minimum is attained, it is attained at a characteristic function. The argument is based on a co-area formula established in [7] by Fleming and Rishel and generalized by Federer:

$$\iint_\Omega |\text{grad } u| \, dx dy = \int_{-\infty}^{\infty} (\text{relative perimeter of } E(s)) \, ds. \tag{6}$$

Here $E(s)$ is the set $\{(x, y) \in \Omega: u(x, y) \geq s\}$. In other words, from the level sets of u we can recover the norm $\|u\| = \iint |\text{grad } u|$, and (6) can be written more concisely as

$$\|u\| = \int_{-\infty}^{\infty} \|\chi_{E(s)}\| \, ds. \tag{7}$$

At the same time the linear functional L satisfies

$$L(u) = \int_{-\infty}^{\infty} L(\chi_{E(s)}) \, ds. \tag{8}$$

Therefore in minimizing the ratio $\|u\|/L(u)$, we can look for the infimum over characteristic functions.

For such a function the ratio is particularly simple. Suppose the area of E is $|E|$, and its perimeter comes from a part of length A in the interior of Ω and a part of length B along the boundary $\partial\Omega$:

$$A = \iint |\text{grad } \chi_E|, \qquad B = \int \chi_E ds.$$

Then substituting $u = \chi_E$ in the ratio gives

$$\cos \gamma_{\text{critical}} = \inf_E \frac{A}{B - \frac{|\Gamma|}{|\Omega|}|E|}. \tag{9}$$

The infimum is taken over all subsets E of Ω for which the denominator is positive; if the sign is negative for some E then it is reversed for the complement $\Omega \backslash E$, and the numerator is unchanged.

The formula (9) is due to Giusti [8]. Earlier Concus and Finn had pointed out that the ratio on the right side always provided an upper bound for the critical cosine; Giusti proved equality at the extremum. His formula is very much simpler than the original minimization over all u, but the calculation can still be delicate—there are many examples in [5] to show how small variations in Ω can produce large variations in the optimal E and γ.

We do know that any part of ∂E interior to Ω must be a circular arc, and for an elliptical Ω that allowed Albright and Doss [1] to compute the optimum. In this example there are only two unknown parameters, one to give the radius of the arc which crosses the ellipse and the other to specify the (symmetric) pair of points at which the arc and the ellipse should meet. (Since the length B comes from an elliptic integral, it is not just an exercise in calculus; there is no simple expression for γ_{critical}.) We note that some argument is required to show that an E of this type—cut off from its complement $\Omega \backslash E$ by a circular arc across Ω—is optimal for a general Ω.

Remark 3. The corresponding result for the norm

$$|||u||| = \iint_{\bar{\Omega}} |\text{grad } u| = \|u\| + \int_{\Gamma} |u| \, ds$$

is due to Fleming [6]. In that case, the extra term represented by the boundary integral means that constant functions are no longer ignored; we have a norm on BV rather than BV/R. The co-area formula now admits the entire perimeter of $E_{(s)}$ rather than discarding its intersection with the boundary Γ. As in [11], the key to optimization over BV is that the extreme points for $|||u||| = 1$ are multiples of characteristic functions χ_E. According to Fleming, ∂E should be a simple closed curve of finite length.

In the applications this integral over $\bar{\Omega}$ arises when the natural boundary condition (the Neumann condition) is changed to a Dirichlet condition $u = 0$ on Γ. It would seem that it is exactly this case in which the boundary term would vanish. But in fact $u = 0$ must be interpreted as a condition only on the external trace, and it is very likely that the physical solution violates $u = 0$ as we approach some part of Γ from the inside. This happened already in the first example [11], and was seen by Temam [12, 13] to be completely typical of a 'relaxed' variational problem. There the original problem

$$\text{Minimize} \quad \iint |\text{grad}\, u|,$$

$$\text{subject to} \quad \iint u = 1 \quad \text{and} \quad u = 0 \text{ on } \Gamma$$

became properly posed only when relaxed to

$$\text{Minimize} \quad |||u|||,$$

$$\text{subject to} \quad \iint u = 1.$$

In the language of plasticity, this means that a displacement boundary condition $u = 0$ cannot be imposed in limit analysis. The fact that the collapse mechanism violates this condition should be taken as normal and in fact the condition does not entirely disappear: It is responsible for the boundary term in the relaxed problem. We are convinced that a minimizing u in that problem deserves to be accepted as a genuine collapse mechanism. This differs from the stricter interpretation adopted in part of the plasticity literature (cf. [10]). It also differs from current practice in finite element calculations; mathematically this boundary condition should not be imposed, but in practice it is.

3. Perfect plasticity

We summarize the problem of anti-plane shear, whose simplicity derives from its special geometry. We are given an infinitely long pipe, again with axis in the z-direction and with cross-section Ω. All external forces act along the axis, and everything is independent of z. Therefore the only displacement u is in the z-direction, and the only stresses are the shears $\sigma_1 = \sigma_{xz}(x, y)$ and $\sigma_2 = \sigma_{yz}(x, y)$. The standard equations of equilibrium are

$$\frac{\partial \sigma_1}{\partial x} + \frac{\partial \sigma_2}{\partial y} + \text{body force} = 0 \quad \text{in } \Omega, \tag{10}$$

$$\sigma \cdot n = \text{surface traction on } \Gamma. \tag{11}$$

The latter may be prescribed on a subset Γ_F of the boundary, with displacement $u = 0$ prescribed on the complement Γ_u. Note that these equations and boundary conditions are not at all sufficient to determine σ_1 and σ_2; there remains a divergence-free component, corresponding to zero external forces, which must be fixed either by an additional equation or a minimum principle.

Plasticity implies that the stresses cannot go outside the 'yield surface' in stress space. For a material that hardens as it undergoes strain, this surface can move; for a perfectly plastic material it is fixed. In our case of anti-plane shear the natural condition, with yield stress normalized to unity, is

$$\sigma_1^2 + \sigma_2^2 \leq 1 \quad \text{in } \Omega. \tag{12}$$

Equality holds in the plastic region, and inequality in the elastic region.

The problem of *limit analysis* is to decide when (10) and (11) become incompatible with the constraint (12). We consider a specific case, in which the pipe is being pulled in one direction by a uniform force along its surface, and pulled in the other direction (the negative z-direction) by a uniform force in the interior. The two forces are kept in balance to avoid simple translation of the whole pipe. Therefore a surface force λ, acting along the circumferential length $|\Gamma|$, is matched with a body force $\lambda|\Gamma|/|\Omega|$ across the cross-sectional area $|\Omega|$. The problem is to discover at what value of λ the resulting stresses lead to collapse. As long as there exists an acceptable stress to balance these external forces, the pipe is safe. Ultimately, however, λ reaches a limit at which the pipe collapses. This limit comes from a convex optimization problem with variables λ, σ_1, σ_2:

$$\text{(P)} \qquad \begin{aligned} &\text{Maximize} \quad \lambda, \\ &\text{subject to} \quad \sigma_1^2 + \sigma_2^2 \leq 1 \quad \text{in } \Omega, \\ &\qquad \frac{\partial \sigma_1}{\partial x} + \frac{\partial \sigma_2}{\partial y} = \lambda \frac{|\Gamma|}{|\Omega|} \quad \text{in } \Omega, \\ &\qquad \sigma \cdot n = \lambda \quad \text{on } \Gamma. \end{aligned}$$

We do not know the stress distribution σ_1, σ_2 at this limit. (Almost certainly it is not uniquely determined.) But we can compute the optimal λ, by duality. To reach the dual we rewrite the primal problem above as

$$\text{(L)} \qquad \sup_{\sigma_1^2+\sigma_2^2 \leq 1} \ \inf_{L(u)=1} L'(u).$$

Here $L(u)$ is the same linear functional as in capillarity, and L' is new:

$$L(u) = \int u \, ds - \frac{|\Gamma|}{|\Omega|} \iint u \, dx dy,$$

$$L'(u) = \int u\sigma \cdot n \, ds - \iint u \operatorname{div} \sigma \, dx dy.$$

To see that the Lagrangian problem (L) is equivalent to (P), we compute the infimum. If σ satisfies the two equations in (P), then $L'(u)$ is just $\lambda L(u)$—whose infimum subject to $L(u) = 1$ is just λ. If on the other hand σ violates one or both of these equations, the infimum is $-\infty$; a linear functional L' is constrained by L only if L' is a multiple of L. The supremum in the Lagrangian will therefore ignore all σ which lead to $-\infty$, and consider only those which satisfy $\operatorname{div} \sigma = \lambda|\Gamma|/|\Omega|$ and $\sigma \cdot n = \lambda$ for some multiple λ. Then the supremum maximizes this multiple and we are back to the primal.

To go from the Lagrangian toward the dual, we interchange sup and inf. At the same time we notice that integration by parts, or Green's theorem, gives L' the simpler form $\iint \sigma \cdot \text{grad}\, u \,\text{d}x\text{d}y$. The intermediate result is therefore

$$\inf_{L(u)=1} \sup_{\sigma_1^2+\sigma_2^2\le 1} \iint \sigma \cdot \text{grad}\, u \,\text{d}x\text{d}y.$$

Now we eliminate σ by computing the supremum. The integrand is largest when σ is in the direction of grad u; because of the restriction on its length, this means that σ is the unit vector grad $u/|\text{grad}\, u|$. With this substitution the integrand is $|\text{grad}\, u|^2/|\text{grad}\, u|$, and we reach the dual:

$$\text{(D)} \qquad \inf_{L(u)=1} \iint |\text{grad}\, u| \,\text{d}x\text{d}y.$$

This is the coincidence, that *the dual problem for antiplane shear leads to the same minimization* (3) *as capillarity without gravity*. If the minimum is attained by a characteristic function $u = \chi_E$, then in both cases the boundary ∂E has a physical interpretation. For plasticity it represents the "hinge line" across which the pipe will shear—for any load above the limit the pipe will break, and the part with cross-section E moves away from the complement. For capillarity this line may be a kind of internal wall, separating the part of the surface which remains finite from the part which climbs to infinity.

4. Extension to other norms

Up to now we have accepted the Euclidean definition $(u_x^2 + u_y^2)^{1/2}$ for the quantity $|\text{grad}\, u|$. It is correct both for capillarity and for the standard laws of plasticity, but still it is not the only possibility: For a crystalline material the constraint $\sigma_1^2 + \sigma_2^2 \le 1$ on the stresses might separate into $|\sigma_1| \le 1$ and $|\sigma_2| \le 1$. This must bring a corresponding change in the dual problem, and it becomes

$$\text{(D}'\text{)} \qquad \begin{aligned} &\text{Minimize} \quad \iint_{\Omega} \left|\frac{\partial u}{\partial x}\right| + \left|\frac{\partial u}{\partial y}\right|, \\ &\text{subject to} \quad L(u) = 1. \end{aligned}$$

In general, each choice of norm for σ will lead to the dual norm or 'polar norm' for grad u. By making this change at every point of Ω we arrive at a new and equivalent norm for BV; the integral in (D′) is finite if and only if the integral in (D) is finite. However it remains to ask whether characteristic functions are still minimizing.

For this we need to extend the co-area formula. The first step is to recompute the norm of a characteristic function, in other words to define the perimeter of a set with respect to an arbitrary norm on R^2. In our example this new perimeter will be

$$\|\chi_E\|' = \iint \left|\frac{\partial \chi_E}{\partial x}\right| + \left|\frac{\partial \chi_E}{\partial y}\right| = \int |n_x| + |n_y|.$$

In the general case it is similar: *The perimeter of E is the integral of the new norm* $|n|'$ *of the unit normal vector.* For the Euclidean case this norm equals one (by definition) and we have the usual perimeter of E—or relative perimeter, if the space is BV/R and there are no contributions from the boundary.

Federer has shown us how the co-area formula extends to this change of norm. The general formula is

$$\iint_\Omega v(x, y)|\text{grad}\, u| = \int_{-\infty}^{\infty} \int_{E(s)} v(x, y)\mathrm{d}H^1 \mathrm{d}s,$$

where v is an arbitrary L^1 function and H^1 is Hausdorff measure. For our application we take $v = |\text{grad}\, u|'/|\text{grad}\, u|$. The formula then finds the integral of the new norm of grad u from the new perimeters of the level sets of u:

$$\iint |\text{grad}\, u|' = \int_{-\infty}^{\infty} \|\chi_{E(s)}\|' \mathrm{d}s.$$

It follows as before that the characteristic functions of Caccioppoli sets are minimizing in our variational problem.

References

[1] N. Albright and S. Doss, "Some properties of capillary surfaces on elliptical domains", to appear.

[2] P. Concus and R. Finn, "On the behavior of a capillary surface in a wedge", *Proceedings of the National Academy of Sciences* 63 (1969) 292–299.

[3] P. Concus and R. Finn, "On capillary free surfaces in the absence of gravity", *Acta Mathematica* 132 (1974) 177–198.

[4] H. Federer, *Geometric measure theory*, Springer-Verlag, New York (1969).

[5] R. Finn, "Existence and nonexistence of capillary surfaces", *Manuscripta Mathematica* 28 (1979) 1–11.

[6] W. Fleming, "Functions with generalized gradient and generalized surfaces", *Annali di Matematica* 44 (1954) 93–103.

[7] W. Fleming and R. Rishel, "An integral formula for total gradient variation", *Archiv der Mathematik* 11 (1960) 218–222.

[8] E. Giusti, "Boundary value problems for nonparametric surfaces of prescribed mean curvature", *Annali della Scuola Normale Superiore di Pisa* 3 (1976) 501–548.

[9] E. Giusti, "On the equation of surfaces of prescribed mean curvature-existence and uniqueness without boundary condition", *Inventiones Mathematicae* 46 (1978) 111–137.

[10] E.M. Shoemaker, "On nonexistence of collapse solutions in rigid-perfect plasticity", *Utilitas Mathematicae* 16 (1979) 3–13.

[11] G. Strang, "A minimax problem in plasticity theory", in: M.Z. Nashed, ed., *Functional analysis*

methods in numerical analysis, Lecture Notes in Mathematics 701, Springer–Verlag, New York (1979) pp. 319–333.

[12] R. Temam, "Mathematical problems in plasticity theory", in: R.W. Cottle, F. Giannessi and J.-L. Lions, eds., *Variational inequalities and complementarity problems*, Wiley, New York (1980) pp. 357–373.

[13] R. Temam and G. Strang, "Functions of bounded deformation", *Archive for Rational Mechanics and Analysis* 75 (1980) 7–21.

Mathematical Programming Study 17 (1982) 103–110.
North-Holland Publishing Company

TIME-DEPENDENT CONTACT PROBLEMS IN RIGID BODY MECHANICS*

Per LÖTSTEDT

Department of Numerical Analysis and Computing Science, Royal Institute of Technology, S-100 44 Stockholm, Sweden

Received 8 September 1980
Revised manuscript received 26 November 1980

Time-dependent contact problems in rigid body mechanics are studied using concepts from linear complementarity theory. Mechanical systems where friction can be neglected are found to be of simpler structure than those exposed to Coulomb friction. A numerical algorithm is outlined for friction-free problems.

Key words: Rigid Body, Mechanical System, Contact Problem, Friction, Linear Complementarity, Quadratic Programming.

1. Introduction

When we wish to study the motion of a mechanical system, a rigid body model often yields sufficiently accurate results. Examples of areas of application are: machinery and mechanisms, biomechanics, road vehicles, trains and satellites. The simulation of such systems on computers is surveyed in [11, 15, 17]. In these papers the motion of the bodies is restricted by bilateral kinematical constraints. Less attention has been paid to systems described by unilateral constraints such as contact problems, see [3, 14] for examples. Related topics for time-independent problems in linear elasticity are treated e.g. in [7, 8]. Dry friction between two bodies sliding on each other cannot always be neglected. A simple, but not completely satisfactory law is attributed to Coulomb, see [10].

2. Frictionfree contacts

Let $q \in R^m$ be the vector of coordinates of the system. In addition to satisfying the equations of motion, q must also satisfy

$$\phi_i(q) \geq 0, \quad i = 1, 2, \ldots, p, \tag{2.1}$$

where ϕ_i, e.g., represents the distance between a corner of one body and the

*This research has obtained financial support from The Swedish Institute of Applied Mathematics and The National Swedish Board for Technical Development.

edge or the surface of another body. Define the set

$$J_N = \{i \mid \phi_i = 0\}. \tag{2.2}$$

In order to fulfil $\phi_i = 0$ for $i \in J_N$ in (t_0, t_1) the Lagrange multiplier terms $g_i\lambda_i$, $i \in J_N$, $g_i = \partial\phi_i/\partial q \in R^m$, $\lambda_i \geq 0$, $\lambda_i \in R$, are added to the equations of motion [10, 19], yielding a system of ordinary differential equations to be solved for $q(t)$ in (t_0, t_1)

$$M(q)\ddot{q} = f(q, \dot{q}, t) + G(q)\lambda. \tag{2.3}$$

$M \in R^{m\times m}$ is symmetric, positive definite, $f \in R^m$ contains driving and inertia forces, $G = (g_{i_1}, g_{i_2}, \ldots, g_{i_n})$ and $\lambda^T = (\lambda_{i_1}, \lambda_{i_2}, \ldots, \lambda_{i_n})$, $i_j \in J_N$. λ_i is proportional to the contact force in the normal direction. A dot denotes a time-derivative and the superscript T a transposition. ϕ_i and λ_i satisfy the following complementarity condition

$$\phi_i \geq 0, \qquad \lambda_i \geq 0, \qquad \lambda_i\phi_i = 0. \tag{2.4}$$

Assume that the first and second derivatives of ϕ_i are continuous. Since $\phi_i = 0$, $i \in J_N$ in (t_0, t_1), we also have

$$\frac{\mathrm{d}\phi_i}{\mathrm{d}t} = g_i^{\mathrm{T}}\dot{q} = 0, \tag{2.5}$$

$$\delta_i = \frac{\mathrm{d}^2\phi_i}{\mathrm{d}t^2} = g_i^{\mathrm{T}}\ddot{q} + \dot{g}_i^{\mathrm{T}}\dot{q} = 0,$$

$$\delta_i^{\mathrm{T}} = (\delta_{i_1}, \delta_{i_2}, \ldots, \delta_{i_n}), \quad i_j \in J_N. \tag{2.6}$$

By introducing $\ddot{q}$ from (2.3) into (2.6), the equation for λ in (t_0, t_1) is

$$G^{\mathrm{T}}M^{-1}G\lambda + G^{\mathrm{T}}M^{-1}f + \dot{G}^{\mathrm{T}}\dot{q} = 0. \tag{2.7}$$

If $G^{\mathrm{T}}M^{-1}G$ is non-singular, then (2.7) can be solved for λ

$$\lambda = -(G^{\mathrm{T}}M^{-1}G)^{-1}(G^{\mathrm{T}}M^{-1}f + \dot{G}^{\mathrm{T}}\dot{q}). \tag{2.8}$$

Insert λ into (2.3). The resulting equation

$$\ddot{q} = (M^{-1} - M^{-1}G(G^{\mathrm{T}}M^{-1}G)^{-1}G^{\mathrm{T}}M^{-1})f - M^{-1}G(G^{\mathrm{T}}M^{-1}G)^{-1}\dot{G}^{\mathrm{T}}\dot{q} \tag{2.9}$$

can be analysed using the well-known theory for ordinary differential equations [1].

Suppose that a constraint j is added to J_N at t_0, i.e., $\phi_j > 0$, $t < t_0$, but $\phi_j = 0$, $t \geq t_0$, a collision has occurred between two bodies. Introduce the notation $\dot{q}^1 = \dot{q}(t_0 - 0)$ and $\dot{q}^2 = \dot{q}(t_0 + 0)$. In general, $\dot{q}$ is discontinuous at t_0 since $g_j^{\mathrm{T}}\dot{q}^1 < 0$ but $g_j^{\mathrm{T}}\dot{q}^2 \geq 0$. A simple jump condition for $\dot{q}$ is

$$M(\dot{q}^2 - \dot{q}^1) = G\Lambda,$$

$$G^{\mathrm{T}}\dot{q}^2 \geq 0, \qquad \Lambda \geq 0, \qquad \Lambda^{\mathrm{T}}G^{\mathrm{T}}\dot{q}^2 = 0, \tag{2.10}$$

[10], a linear complementarity problem (LCP). G in (2.10) consists of all g_i such

that $i \in J_N$ at t_0. Ingleton [9] proves that $G\Lambda$ and $\dot{q}^2$ are unique and if G has full column rank, then Λ is also unique. The components of Λ are proportional to impulses in the mechanical system.

Suppose that a constraint k is dropped from J_N at t_1, i.e., $\phi_k = 0$, $t \leq t_1$, but $\phi_k > 0$, $t > t_1$. Since

$$\phi_k(t) = g_k^T \dot{q}(t)(t - t_1) + O((t - t_1)^2)$$

it follows from (2.4) that there is a $\Delta t' > 0$ such that

$$g_k^T \dot{q} \geq 0, \qquad \lambda_k g_k^T \dot{q} = 0, \qquad t \in (t_0, t_1 + \Delta t']. \tag{2.11}$$

Since

$$g_k^T \dot{q}(t) = \delta_k(t)(t - t_1) + O((t - t_1)^2)$$

there is a Δt, $0 < \Delta t < \Delta t'$ such that

$$\delta_k \geq 0, \qquad \lambda_k \delta_k = 0, \qquad t \in (t_0, t_1 + \Delta t]. \tag{2.12}$$

Hence, the governing equations and relations in an interval $(t_0, t_1 + \Delta t]$, where constraints are dropped from J_N, are (2.3) and

$$G^T \ddot{q} + \dot{G}^T \dot{q} = \delta \geq 0 \tag{2.13}$$

or after replacing $\ddot{q}$ in (2.13) by $\ddot{q}$ from (2.3),

$$\begin{aligned} &G^T M^{-1} G\lambda + G^T M^{-1} f + \dot{G}^T \dot{q} = \delta \geq 0 \\ &\lambda \geq 0, \qquad \lambda^T \delta = 0, \end{aligned} \tag{2.14}$$

a time-dependent LCP. G is determined by J_N in (t_0, t_1). Since $\lambda_k = 0$ in $[t_1, t_1 + \Delta t']$, $g_k \lambda_k$ will not affect the solution of (2.3) in that interval.

After a discontinuity at t_0 or initially ($t_0 = 0$), those constraints k for which $g_k^T \dot{q}(t_0) > 0$ can be dropped immediately from J_N for $t > t_0$. This is shown by considering the Taylor expansion for ϕ_k about t_0. Solve (2.14) for $\lambda(t_0)$ and $\delta(t_0)$. Similarly, for a constraint k such that $\delta_k(t_0) > 0$ we have $k \notin J_N$ for $t > t_0$. However, it is incorrect to let J_N for $t > t_0$ be equivalent to $\{i \mid \phi_i(t_0) = g_i^T \dot{q}(t_0) = \delta_i(t_0) = 0\}$, since there may exist a j such that $\lambda_j(t_0) = \delta_j(t_0) = 0$ but $\delta_j(t_0) > 0$ and $j \notin J_N$ for $t > t_0$. Nevertheless, assuming sufficiently smooth M, f and G in (2.3), the result by Ingleton [9] can be applied repeatedly to the time-derivatives of λ and δ to obtain the following theorem.

Theorem 1. *Let $J = \{i \mid \phi_i = 0$ and $g_i^T \dot{q} = 0$ at $t = t_0\}$ and let G consist of those n columns g_i for which $i \in J$. Assume that $M(q)$, $G(q)$ and $f(q, \dot{q}, t)$ are analytic in t and all q_i, $\dot{q}_i$ separately, and that G, f and $\partial G / \partial q$ are bounded in the neighbourhood of $q(t_0)$, $\dot{q}(t_0)$ and t_0. Furthermore, assume that the rank of G is constant in the neighbourhood of $q(t_0)$.*

Then J_N is constant in $(t_0, t_0 + \Delta t)$ for a sufficiently small Δt. The solution $q(t)$ of (2.3) is analytic and unique in $[t_0, t_0 + \Delta t)$. If G has full

full column rank, $\lambda(t)$ is also unique and analytic. There is a set $J' \subset J$ of cardinality $r \leq rank(G)$ such that $G' = (g_{i_1}, g_{i_2}, \ldots, g_{i_r})$, $i_j \in J'$, $rank(G') = r$, and there is a $\lambda'(t) \in R^r$, $\lambda_i'(t) > 0$ in $(t_0, t_0 + \Delta t)$, *such that $G\lambda(t) = G'\lambda'(t)$.*

Proof. The proof in [12] for a constant M is easily modified to take $M(q)$ into account.

When J_N in $(t_0, t_0 + \Delta t)$ has been determined, the solution can also be computed by (2.9) with $G = G'$. $\lambda'(t)$ is given by (2.8).

By an accumulation point is here meant a point t_2 in the neighbourhood of which the number of changes of J_N is infinite. Near such a point Δt in the theorem is necessarily small. The point where a bouncing ball finally lies at rest is an example of such a point. When the solution approaches t_2, Δt decreases geometrically in this example.

3. Coulomb friction

We restrict the discussion of Coulomb friction to planar problems. Let

$$g_{Fi}^T \dot{q}, \quad i = 1, 2, \ldots, l \leq p, \tag{3.1}$$

be the relative velocity in the tangential direction at a contact between two bodies. In order for $q(t)$ to satisfy $g_{Fi}^T \dot{q} = 0$, there is a Lagrange multiplier term $g_{Fi}\Lambda_i$ in the equations of motion (2.3). The friction multiplier Λ_i satisfies $|\Lambda_i| \leq \mu\lambda_i$. μ is the constant, positive friction coefficient and λ_i is the normal multiplier at the same contact point. When one body is sliding on the other body, then $g_{Fi}^T \dot{q} \neq 0$ and $|\Lambda_i| = \mu\lambda_i$, $\text{sign}(\Lambda_i) = -\text{sign}(g_{Fi}^T \dot{q})$. Let J_{F0} and J_{FW} be defined by

$$J_{F0} = \{i \mid i \in J_N,\ g_{Fi}^T \dot{q} = 0\},$$
$$J_{FW} = \{i \mid i \in J_N,\ g_{Fi}^T \dot{q} \neq 0\}. \tag{3.2}$$

Hence, we have

$$\Lambda_i g_{Fi}^T \dot{q} \leq 0, \qquad |\Lambda_i| = \mu\lambda_i, \quad i \in J_{FW}. \tag{3.3}$$

Suppose that a constraint k is dropped from J_{F0} and added to J_{FW} at t_1. Introduce $\Delta_k = g_{Fk}^T \ddot{q} + \dot{g}_{Fk}^T \dot{q}$ and use arguments similar to those leading to (2.12) to obtain relations between Λ_k and Δ_k:

$$-\mu\lambda_k \leq \Lambda_k \leq \mu\lambda_k, \qquad t \in (t_0, t_1 + \Delta t] \quad \text{for some } \Delta t > 0,$$
$$\text{if } \Delta_k \neq 0, \quad \text{then } |\Lambda_k| = \mu\lambda_k \quad \text{and} \quad \Lambda_k\Delta_k \leq 0. \tag{3.4}$$

A constraint k is dropped from J_{FW} and added to J_{F0} when a previously non-zero $g_{Fk}^T \dot{q}$ vanishes. Then $\ddot{q}$ is in general discontinuous. When a constraint is dropped from J_N, then an associated friction constraint is dropped from J_{F0} or J_{FW}.

The notation is simplified if we assume that the constraints are ordered such that $J_N = \{1, 2, \ldots, n\}$, $J_{F0} = \{1, 2, \ldots, s\}$ and $J_{FW} = \{s+1, s+2, \ldots, r\}$, $r \le n$. Let

$$\begin{aligned} G_* &= (g_1, \ldots, g_n, g_{F1}, \ldots, g_{Fs}), \\ H &= (g_{F,s+1}, \ldots, g_{Fr}), \\ \lambda_*^T &= (\lambda_1, \ldots, \lambda_n, \Lambda_1, \ldots, \Lambda_s), \\ \delta_*^T &= (\delta_1, \ldots, \delta_n, \Delta_1, \ldots, \Delta_s). \end{aligned} \tag{3.5}$$

There is a $U \in R^{(r-s)\times(n+s)}$ such that the contribution of the working friction forces (i.e., Λ_i, $i \in J_{FW}$) to the equations of motion can be written

$$\sum_{i\in J_{FW}} g_{Fi}\Lambda_i = \sum_{i\in J_{FW}} g_{Fi}s_i\mu\lambda_i = HU\lambda_* \tag{3.6}$$

where $s_i = -\operatorname{sign}(g_{Fi}^T\dot{q})$. The equations of motion are (cf. (2.3), (3.5), (3.6))

$$M\ddot{q} = f + (G_* + HU)\lambda_*. \tag{3.7}$$

The friction counterpart of (2.14) in an interval $(t_0, t_1 + \Delta t]$, where constraints are dropped from J_N and J_{F0}, is obtained by introducing $b = G_*^T M^{-1} f + \dot{G}_*^T\dot{q}$ and combining (2.12), (3.4) and (3.7)

$$\begin{aligned} &G_*^T M^{-1}(G_* + HU)\lambda_* + b = \delta_*, \\ &\lambda_i \ge 0, \quad \delta_i \ge 0, \quad \lambda_i\delta_i = 0, \\ &-\mu\lambda_i \le \Lambda_i \le \mu\lambda_i, \\ &\text{if } \Delta_i \ne 0, \text{ then } |\Lambda_i| = \mu\lambda_i \text{ and } \Lambda_i\Delta_i \le 0. \end{aligned} \tag{3.8}$$

G_* and H are defined by J_N, J_{F0} and J_{FW} in (t_0, t_1) when the sets are constant.

The equations for λ_* and $\ddot{q}$ in (t_0, t_1) when friction is included, corresponding to (2.8) and (2.9), are

$$\lambda_* = -(G_*^T M^{-1}(G_* + HU))^{-1}(G_*^T M^{-1} f + \dot{G}_*^T\dot{q}), \tag{3.9}$$

$$\begin{aligned} \ddot{q} = {} & (M^{-1} - M^{-1}(G_* + HU)(G_*^T M^{-1}(G_* + HU))^{-1} G_*^T M^{-1})f \\ & - M^{-1}(G_* + HU)(G_*^T M^{-1}(G_* + HU))^{-1}\dot{G}_*^T\dot{q}. \end{aligned} \tag{3.10}$$

To determine J_N, J_{F0} and J_{FW} for $t > t_0$ after a discontinuity at t_0 or initially, then drop the constraints i, $i \in J_N$ at t_0, for which $g_i^T\dot{q}(t_0) > 0$ and solve (3.8). The columns of G are g_i, $i \in J_N' = \{j \mid \phi_j(t_0) = g_j^T\dot{q}(t_0) = 0\}$, and g_{Fi}, $i \in J_{F0}' = J_{F0} \cap J_N'$ at t_0, and those of H are g_{Fi}, $i \in J_{FW}' = J_{FW} \cap J_N'$ at t_0. Simple examples by Painlevé [16] and several other authors reveal that a solution to (3.8) does not always exist. Under the assumption that $J_{F0} = \emptyset$, then $\Lambda_i = s_i\mu\lambda_i$, $i \le r$, and (3.8) is an LCP to solve for λ_i and δ_i,

$$\begin{aligned} &G_*^T M^{-1}(G_* + HU)\lambda_* + b = \delta_* \ge 0, \\ &\lambda_* \ge 0, \quad \lambda_*^T\delta_* = 0. \end{aligned} \tag{3.11}$$

A P-matrix is a matrix whose principal minors all are positive. Cottle and

Dantzig [2] prove that if $G_*^{\mathrm{T}}M^{-1}(G_* + HU)$ is a P-matrix, then there exists a solution λ_*, δ_* to (3.11) for every b. The ideas in [2] can be generalized also to the solution of (3.8). In [12] an algorithm is presented and the following theorem is proved by construction.

Theorem 2. *Let J_{F1} be any subset (including $\emptyset$) of J'_{F0} and $J_{\mathrm{F2}} = J'_{\mathrm{F0}} \setminus J_{\mathrm{F1}}$. Define $G_0 = (g_1, \ldots, g_n, g_{\mathrm{F}i_1}, \ldots, g_{\mathrm{F}i_t})$, $\lambda_0^{\mathrm{T}} = (\lambda_1, \ldots, \lambda_n, \Lambda_{i_1}, \ldots, \Lambda_{i_t})$, $i_j \in J_{\mathrm{F1}}$ and $H_0 = (H, g_{\mathrm{F}i_1}, \ldots, g_{\mathrm{F}i_u})$, $i_j \in J_{\mathrm{F2}}$. U_0 is determined by*

$$\sum_{i \in J_{\mathrm{FW}}} g_{\mathrm{F}i} s_i \mu \lambda_i + \sum_{i \in J_{\mathrm{F2}}} g_{\mathrm{F}i} s_i \mu \lambda_i = H_0 U_0 \lambda_0.$$

If $G_0^{\mathrm{T}} M^{-1}(G_0 + H_0 U_0)$ is a P-matrix for any set J_{F1} with any possible choice of $s_i = \pm 1$, $i \in J_{\mathrm{F2}}$, then there exists a solution λ_*, δ_* to (3.8) at t_0 for every b.

A sufficient condition for the solution to be unique is also given in [12].

Assume that G_* and H fulfil the condition in Theorem 2 and the solution λ_*, δ_* at t_0 is non-degenerate in the following sense

$$\text{if } \delta_i = 0, \quad \text{then } \lambda_i > 0,$$
$$\text{if } \Delta_i = 0, \quad \text{then } \mu\lambda_i - |\Lambda_i| > 0.$$

If $\lambda_*(t)$ in (3.9) and $q(t)$, $\dot{q}(t)$ are continuous, then there is a sufficiently small $\Delta t > 0$ such that J_{N}, J_{F0} and J_{FW} in $(t_0, t_0 + \Delta t)$ are

$$J_{\mathrm{N}} = \{i \mid \phi_i(t_0) = g_i^{\mathrm{T}}\dot{q}(t_0) = \delta_i(t_0) = 0\},$$
$$J_{\mathrm{F0}} = \{i \mid i \in J_{\mathrm{N}},\ g_{\mathrm{F}i}^{\mathrm{T}}\dot{q}(t_0) = \Delta_i(t_0) = 0\},$$
$$J_{\mathrm{FW}} = \{i \mid i \in J_{\mathrm{N}},\ g_{\mathrm{F}i}^{\mathrm{T}}\dot{q}(t_0) \neq 0 \text{ or } \Delta_i(t_0) \neq 0\}.$$

The governing equations of motion in $(t_0, t_0 + \Delta t)$ are (3.10).

4. On the numerical solution of frictionfree systems

The LCP (2.14) has the same solution as the quadratic programming problem, [2],

$$\min_{\lambda \geq 0} \tfrac{1}{2}\lambda^{\mathrm{T}} G^{\mathrm{T}} M^{-1} G \lambda + \lambda^{\mathrm{T}}(G^{\mathrm{T}} M^{-1} f + \dot{G}^{\mathrm{T}} \dot{q}). \tag{4.1}$$

The dual problem of (4.1) taking (2.3) into account is, [5],

$$\min_{\ddot{q}} \tfrac{1}{2}(\ddot{q} - M^{-1}f)^{\mathrm{T}} \mathrm{M}(\ddot{q} - M^{-1}f),$$
$$G^{\mathrm{T}}\ddot{q} + \dot{G}^{\mathrm{T}}\dot{q} \geq 0. \tag{4.2}$$

In (4.2) q and $\dot{q}$ are to be considered as fixed. (4.2) is a generalization of the principle of least constraint by Gauss, [18], to rigid body systems subject to unilateral constraints.

We shall outline a numerical algorithm for solving (4.1), (4.2) in a special case. In a planar system the coordinates can always be chosen such that M is diagonal and constant. Furthermore, if the external forces are independent of $\dot{q}$, then $f = f(q, t)$. Let a subscript i on a variable denote its approximate value at discrete time-points $t_0, t_1, \ldots, t_{i+1} = t_i + h$. Suppose that q_i and $\dot{q}_i$, $i < j$, are known. First q is advanced one time-step to q_j by an explicit linear multistep method, [4],

$$\sum_{i=0}^{k} \alpha_i q_{j-i} = h \sum_{i=1}^{k} \beta_i \dot{q}_{j-i}. \tag{4.3}$$

Replace $\ddot{q} - M^{-1}f$ in (4.2) at t_j by another linear multistep formula

$$\sum_{i=0}^{l} (\alpha'_i \dot{q}_{j-i} - h\beta'_i M^{-1} f(q_{j-i}, t_{j-i})) \tag{4.4}$$

and the linear inequality constraint by $G_j^T \dot{q}_j \geq 0$. Let

$$b_j = -(1/\alpha'_0) \sum_{i=1}^{l} \alpha'_i \dot{q}_{j-i} + (h/\alpha'_0) M^{-1} \sum_{i=0}^{l} \beta'_i f_{j-i}.$$

Then we obtain a weighted linear least squares problem subject to linear inequality constraints to be solved at each time-step for $\dot{q}_j$,

$$\min_{\dot{q}_j} \tfrac{1}{2}(\dot{q}_j - b_j)^T M (\dot{q}_j - b_j)$$

$$G_j^T \dot{q}_j \geq 0. \tag{4.5}$$

The dual problem of (4.5) is

$$\min_{z \geq 0} \tfrac{1}{2} z^T G_j^T M^{-1} G_j z + z^T G_j^T b_j \tag{4.6}$$

where $\dot{q}_j$ and z are related by

$$\dot{q}_j = b_j + M^{-1} G_j z. \tag{4.7}$$

If the method in (4.4) is chosen such that $\beta'_0 = 1$, $\beta'_i = 0$, $i = 1, 2, \ldots, l$, e.g., a backward difference formula, then (4.7) is a discretization of (2.3) with $\lambda_j = (\alpha'_0/h)z$.

Following the ideas of Gill and Murray [6], (4.6) seems to be the most suitable formulation for computation. λ_{j-1} is a feasible initial guess for step j and the new search direction for a minimum can be calculated by using old decompositions of $M^{-1/2}G_i$. In most time-steps $J_{N,j-1} = J_{Nj}$ and only one system of linear equations has to be solved at each step. For more details see [13].

It sometimes occurs in a time-step that $\phi_i(t_{j-1}) > 0$ but $\phi_i(t_j) \leq 0$. Then compute by inverse interpolation a t_*, $t_{j-1} < t_* \leq t_j$, such that $\phi(t_*) = 0$, solve (2.10) for the new velocities and restart the integration from t_*.

Acknowledgement

This paper is extracted from the author's dissertation, which was written under the auspices of professor Germund Dahlquist.

References

[1] E.A. Coddington and N. Levinson, *Theory of ordinary differential equations* (McGraw-Hill, New York, 1955).

[2] R.W. Cottle and G.B. Dantzig, "Complementary pivot theory of mathematical programming", *Linear Algebra and Applications* 1 (1968) 103–125.

[3] P.A. Cundall and O.D.L. Stark, "A discrete model for granular assemblies", *Géotechnique* 1 (1979) 47–65.

[4] G. Dahlquist and Å. Björck, *Numerical methods* (Prentice-Hall, Englewood Cliffs, NJ, 1974).

[5] W.S. Dorn, "Duality in quadratic programming", *Quarterly of Applied Mathematics* 18 (1960) 155–161.

[6] P.E. Gill and W. Murray, "Numerically stable methods for quadratic programming", NPL Report NAC78, National Physical Laboratory (Teddington, 1977).

[7] R. Glowinski, J.L. Lions and R. Trémolières, *Analyse numérique des inéquations variationnelles*, t. 1–2 (Dunod, Paris, 1976).

[8] E. Haug, R. Chanad and K. Pan, "Multibody elastic contact analysis by quadratic programming", *Journal of Optimization Theory and Applications* 21 (1977) 189–198.

[9] A.W. Ingleton, "A problem in linear inequalities", *Proceedings of London Mathematical Society* (3) 16 (1966) 519–536.

[10] C.W. Kilmister and J.E. Reeve, *Rational mechanics* (Longman's, London, 1966).

[11] A.I. King and C.C. Chou, "Mathematical modelling, simulation and experimental testing of biomechanical system crash response", *Journal of Biomechanics* 9 (1976) 301–317.

[12] P. Lötstedt, "Analysis of some difficulties encountered in the simulation of mechanical systems with constraints", TRITA-NA-7914, Department of Numerical Analysis, Royal Institute of Technology (Stockholm, 1979).

[13] P. Lötstedt, "A numerical method for the simulation of mechanical systems with unilateral constraints", TRITA-NA-7920, Department of Numerical Analysis, Royal Institute of Technology (Stockholm, 1979).

[14] P. Lötstedt, "Interactive simulation of the progressive collapse of a building, revisited", TRITA-NA-7921, Department of Numerical Analysis, Royal Institute of Technology (Stockholm, 1979).

[15] K. Magnus, ed., *Dynamics of multibody systems* (Springer, Berlin, 1978).

[16] P. Painlevé, "Sur les lois du frottement de glissement", *Comptes rendus hebdomadaires des séances de l'Academie des Sciences* 121 (1895) 112–115, 140 (1905) 702–707, 141 (1905) 401–405 and 546–552.

[17] B. Paul, "Analytical dynamics of mechanisms—A computer oriented overview", *Mechanisms and Machine Theory* 10 (1975) 481–507.

[18] E.T. Whittaker, *Analytical dynamics*, 4th edition (Dover, New York, 1944).

[19] J. Wittenburg, *Dynamics of systems of rigid bodies* (Teubner, Stuttgart, 1977).

Mathematical Programming Study 17 (1982) 111–125.
North-Holland Publishing Company

ON SOME RECENT ENGINEERING APPLICATIONS OF COMPLEMENTARITY PROBLEMS†

Ikuyo KANEKO*

Department of Industrial Engineering and Department of Computer Sciences, University of Wisconsin-Madison, WI 53706, U.S.A.

Received 15 July 1980
Revised manuscript received 23 February 1981

In this paper we discuss two recent structural engineering applications of mathematical programming. One of them is an optimum design of certain elastic-plastic structures and the other is the analysis of a certain rigid-plastic workhardening adaptation model. Both problems are closely related to the (parametric) linear complementarity and quadratic programming problems. We shall examine some mathematical properties, such as the existence of an optimal solution, and possible solution methods for these problems.

Key words: Complementarity Problem, Engineering Applications, Engineering Plasticity, Quadratic Programming, Optimum Design Problem.

1. Introduction

In this paper we discuss two recent structural engineering applications of mathematical programming, in particular, complementarity problems. One of them (Problem A) is an optimal design of elastic-plastic structures (c.f. [3]) and the other (Problem B) the analysis of rigid-plastic structures using the concept of work-hardening adaptation (c.f. [6]).

Problem A leads to an optimization problem in which the feasibility is defined as the satisfaction of a system of linear inequalities by the solution of a *linear complementarity problem* (LCP) whose data are (nonlinear) functions of the problem variables. This problem may be solved by an iterative procedure where a 'linear program with a complementarity constraint' is solved in each iteration. Problem B gives rise to a nonlinear programming problem which is closely related to a parametric *quadratic program* (QP). This problem can be solved by using a parametric LCP technique.

The purposes of this paper are to study some mathematical properties (such as the existence of a solution) of Problems A and B, and to examine ways of solving these problems. In this paper we do not intend to analyze or examine the structural engineering aspect. More detailed descriptions of Problems A and B along with their validity and significance as structural engineering problems can

† This research is supported, in part, by a grant from the Graduate School, University of Wisconsin-Madison.

* Visiting professor, University of Essen, West Germany.

be found in papers in the structural mechanics literature, such as [3, 4, 6]. In particular, [3] deals with Problem A from the structural engineering point of view referring to the mathematical results obtained in this present paper. Another paper by the author (with Polizzotto and Mazzarella) is under preparation which examines the structural aspect of Problem B.

To the author's knowledge, Problems A and B as mathematical programs have not been investigated in the literature.

Problems A and B which are formally stated in the next section will be investigated in Sections 3 and 4, respectively. In Appendix 1, we shall demonstrate a certain relationship between the two problems, and in Appendix 2, some examples are presented to illustrate a certain scheme related to the proposed solution method for Problem A.

2. Statement of problems

Maier has shown [4] that one can perform an analysis of a certain class of *elastic-plastic* structures (such as reinforced concrete frames) by solving the LCP

$$q + Mz \geq 0, \qquad z \geq 0, \qquad z^{\mathrm{T}}(q + Mz) = 0, \tag{2.1}$$

where q is a given n-vector and M is a given n by n positive definite matrix. In some cases, the following additional conditions must be considered:

$$Az \leq a. \tag{2.2}$$

Here, A is t by n and a is a t-vector; and (2.2) represents the requirement that the amounts of displacements be within given limits.

By the property of M, the LCP (2.1) can be solved by a standard complementary pivoting algorithm; also, because the LCP has a unique solution, the addition of (2.2) would merely require checking whether or not the solution of the LCP satisfies the constraints. Having found the solution z satisfying all the conditions in (2.1) (and (2.2)) one can then compute the corresponding vectors of *stresses* and *strains* which describe the behavior of the structure [4].

This 'Maier's LCP approach' is to analyze a given structure. Problem A is concerned with an optimal design of a structure, using the same approach, to minimize the cost (weight, etc.) of the structure. Let x be the r-vector of 'independent' design variables.[1] For instance, x_j may represent the area of a cross section of a steel bar element in the structure. The design variables are required to be bounded from below by a positive r-vector, say h.

It is assumed that the cost of the design is a linear function, $c^{\mathrm{T}}x$, of the design variables where c is a positive r-vector. The data, q, M, A and a in (2.1)–(2.2)

[1] Independent design variables are those whose values can be chosen independently of one another.

are in general dependent on x (as the thickness of steel bars affects the strength of the structure, for example), and we shall write $q(x)$, $M(x)$, $A(x)$ and $a(x)$ to indicate the dependencies. Problem A is thus given by

$$\text{minimize } c^{\mathrm{T}}x, \tag{2.3a}$$

$$\text{subject to } \quad q(x) + M(x)z \geq 0, \quad z \geq 0, \tag{2.3b}$$

$$z^{\mathrm{T}}(q(x) + M(x)z) = 0, \tag{2.3c}$$

$$A(x)z \leq a(x), \tag{2.3d}$$

$$x \geq h. \tag{2.3e}$$

In the following we assume that $M(x)$ is positive definite for every $x \geq h$: the resulting uniqueness of the solution of the LCP (2.3b)–(2.3c) enables us to rephrase (2.3) in a way which is more convenient in the analysis of the problem. Let us use the symbol (q, M) to denote the LCP (2.1). Then we have

Problem A.

$$\min \left\{ c^{\mathrm{T}}x: \begin{array}{l} x \geq h \\ \text{the solution } z \text{ of } (q(x), M(x)) \\ \text{satisfies } A(x)z \leq a(x) \end{array} \right\}. \tag{2.4}$$

Problem B is concerned with the determination of the best upper bound on the amount of a displacement or other quantity ("deformation parameter") critical to the structural safety, generated during a (dynamic) loading process. Using the concept of workhardening adaptation, Polizzoto and his co-workers showed (see [6]) that by solving the following nonlinear program one can obtaine such a bound for a class of rigid-plastic structures:

Problem B.

$$\text{minimize } \quad \frac{1}{\omega}(\tfrac{1}{2}z^{\mathrm{T}}Hz + K_0), \tag{2.5a}$$

$$\text{subject to } \quad Hz - N^{\mathrm{T}}y + k - \omega d \geq 0, \tag{2.5b}$$

$$C^{\mathrm{T}}y = f, \tag{2.5c}$$

$$\omega > 0, \quad z \geq 0. \tag{2.5d}$$

The variables in this problem are: z is an n-vector, ω is a scalar, and y is an m-vector. The problem data are: H is an n by n symmetric positive semi-definite *workhardening matrix*, K_0 is a nonnegative scalar, N^{T} and C^{T} are n by m and t by m matrices respectively, k and d are n-vectors and f is a t-vector. The optimal value provides the most stringent upper bound of the value of the relevant deformation parameter generated during the loading process.

By introducing design variables and the problem data which are functions of the design variables, one could consider an optimal design problem based on Problem B (minimizing the cost of the design subject to a given upper limit on the value of the deformation parameter, [6]); but we shall not treat it in this paper.

In the following, we shall touch upon some relationships between Problems A and B. First, the two engineering problems which motivated these problems are different in certain fundamental ways from the structural analysis point of view. For instance, Problem A deals with a 'holonomic' (i.e., time independent problem for elastic-plastic structures, whereas Problem B is a holonomic problem giving a bound for a dynamic problem for rigid-plastic structures. Roughly speaking, the difference between elastic-plastic and rigid-plastic structures is that in the former, elastic as well as plastic strains are considered, while in the latter, only plastic ones are assumed to be present. Despite these differences, the two problems do share a common, underlying structural principle from which both are derived. In fact, the set of formulas describing the mechanical properties of the elastic-plastic structures, which lead to the LCP (2.1), reduces to the Kuhn–Tucker optimality conditions for the nonlinear program (2.5) when specialized to the rigid-plastic case. This will be shown in Appendix 1. The variable vector z in both (2.3) and (2.5) denote the same quantities, the plastic intensities, in the respective underlying structural problems. The variable y (stress vector) in (2.5) is eliminated in (2.3) by using a certain condition particular to Problem A (i.e., a condition not present in Problem B). The scalar variable ω in Problem B is known as the 'perturbation multiplier' which is inserted to the problem in such a way that the optimal value of the problem (2.5) would provide the desired upper bound. These relationships are explained, in more detail, in [6].

3. Problem A

In this section we shall examine some properties and a possible method of solution of Problem A (2.4). Throughout the section we shall assume that the problem satisfies the following:

(1) $q(x) = p(x) + Rx$, where p is a nonlinear function of x such that $p(x) \in \mathbf{R}^n$ and R is an n by r matrix.

(2) p, M, A and a are continuous in x.

(3) $c > 0$.

(4) $M(x)$ is positive definite for every $x \geq h$.

(5) The system (2.3b)–(2.3e) is feasible.

All of these conditions are satisfied in the underlying structural engineering application which motivated Problem A [3].

In the following we shall let

$$X = \left\{ x \in \mathbf{R}^r: \begin{array}{l} x \geq h \\ \text{the solution } z \text{ of the LCP} \\ (p(x) + Rx, M(x)) \text{ satisfies } A(x)z \leq a(x) \end{array} \right\}.$$

Problem A then can be written as:

$$\min\{c^{\mathrm{T}}x: x \in X\}. \tag{3.1}$$

We state, as a lemma, a property of the LCP which will be used frequently in what follows.

Lemma. *Let $\mathcal{M}$ be the class of n by n positive definite matrices. If we let $z(q, M)$ denote the unique solution of the LCP (q, M), then $z(q, M)$ is continuous in $q \in \mathbf{R}^n$ and $M \in \mathcal{M}$.*

Our first result is:

Theorem 1. *Problem* A *has an optimal solution.*

Proof. Let x^0 be a feasible solution of (2.4) which is assumed to exist. Let X^0 be defined by

$$X^0 = X \cap \{x: c^{\mathrm{T}}x \leq c^{\mathrm{T}}x^0\}.$$

Clearly, Problem A is equivalent to minimizing $c^{\mathrm{T}}x$ over X^0. The existence of an optimal solution follows from the compactness of X^0. The set X^0 is clearly bounded; its closedness is a result of the continuity of p, M, A and a, and the continuity of the solution of the LCP $(q(x), M(x))$ in its problem data.

The optimal solution need not be unique as the following example demonstrates.

Example. Consider the problem (3.1) with $n = 2$, $c^{\mathrm{T}} = (1, 1)$, $h^{\mathrm{T}} = (0.1, 0.1)$ $p(x)^{\mathrm{T}} = (x_2 - 1, x_1 - 1)$, $a(x) = 0$ and each of R, $M(x)$ and $A(x)$ is the 2 by 2 identity matrix. Then, any x with $x \geq h$ and $x_1 + x_2 = 1$ is an optimal solution.

There doesn't seem to be an existing algorithm for solving Problem A in a direct and effective way, unless the problem data are independent of x. If p, M, A and a are constant, then the problem is a 'linear program with a complementarity constraint' [2] and can be solved by an existing method such as the branch-and-bound algorithm given in [2]. In a general case, we shall consider an iterative scheme which is described below[2]. This scheme is not ideal in several

[2] The basic idea for this procedure was originally suggested to the author by Maier [5]; the procedure was proposed in [3] to solve the specific engineering problem.

ways (for instance, the convergence is not guaranteed, and it may require much computation to solve large size problems); hopefully, the proposed procedure will motivate the development of more complete solution methods.

For a given $\bar{x} \in X$ and a given positive n-vector γ we denote by $P(\bar{x}; \gamma)$ the following problem:

$$\min\left\{ c^{\mathrm{T}}x: \begin{array}{l} x \geq h, \quad -\gamma \leq x - \bar{x} \leq \gamma \\ \text{the solution } z \text{ of the LCP } (p(\bar{x}) + Rx, M(\bar{x})) \\ \text{satisfies } A(\bar{x})z \leq a(\bar{x}) + \epsilon \end{array} \right\},$$

where ϵ is a fixed t-vector. All through the algorithm we keep $\gamma > \underline{\gamma}$, where $\underline{\gamma}$ is a fixed positive r-vector. A general iteration step of the proposed procedure can be described as follows:

Let $\bar{x} \in X$ and $\gamma > \underline{\gamma}$ be given.

Optimization phase: Solve $P(\bar{x}; \gamma)$. Let $\hat{x}$ be an optimal solution. Terminate the procedure if $\hat{x} = \bar{x}$; otherwise go to the next phase.

Feasibility phase: Determine whether or not $\hat{x} \in X$. If $\hat{x} \in X$, then replace $\bar{x}$ with $\hat{x}$ and start the next iteration, after changing γ (but keep $\gamma > \underline{\gamma}$) if desired. Otherwise, retain $\bar{x}$, reduce the values of all or some components of γ (keeping $\gamma > \underline{\gamma}$) and go to the next iteration.

Remarks. (i) Given $\bar{x} \in X$ and $\gamma > 0$, the problem $P(\bar{x}; \gamma)$ is always feasible ($\bar{x}$ itself is feasible) and has an optimal solution (c.f. the proof of Theorem 1). Clearly, $c^{\mathrm{T}}\hat{x} \leq c^{\mathrm{T}}\bar{x}$.

(ii) Both of the two phases in the procedure can be performed by solving the following 'linear probrams with a complementarity constraint', using the Kaneko–Hallman algorithm mentioned above:

Optimization phase:

$$\min\left\{ c^{\mathrm{T}}x: \begin{array}{l} x \geq h, \quad -\gamma \leq x - \bar{x} \leq \gamma \\ p(\bar{x}) + Rx + M(\bar{x})z \geq 0, \quad z \geq 0 \\ z^{\mathrm{T}}(p(\bar{x}) + Rx + M(\bar{x})z) = 0, \quad A(\bar{x})z \leq a(\bar{x}) + \epsilon \end{array} \right\}.$$

Feasibility phase:

$$\min\left\{ e^{\mathrm{T}}z: \begin{array}{l} p(\bar{x}) + R\bar{x} + M(\bar{x})z \geq 0, \quad z \geq 0 \\ z^{\mathrm{T}}(p(\bar{x}) + R\bar{x} + M(\bar{x})z) = 0 \\ A(\bar{x})z \leq a(\bar{x}) \end{array} \right\},$$

where e is the n-vector of ones. The feasibility phase can also be executed by solving the LCP $(p(\bar{x}) + R\bar{x}, M(\bar{x}))$ and checking the satisfaction of $A(\bar{x})z \leq a(\bar{x})$.

In the following, we shall explain some motivation for the above algorithm. First of all, let us observe that the critical difficulty of the problem (2.3) lies in the nonlinearity of the data q, M, A and a in the variable x. Thus, the basic idea is to use some given $\bar{x}$ to 'fix' the data (i.e., use $q(\bar{x})$, $M(\bar{x})$, $A(\bar{x})$ and $a(\bar{x})$ in (2.3b)–(2.3d)) so that the problem can be solved by an existing algorithm. The obvious difficulty is that an optimal, or even just a feasible solution, say x^*, of such a problem (where the data are fixed by $\bar{x}$) may not be feasible in (2.3) since x^* must satisfy (2.3b)–(2.3e) with $q(x^*)$, $M(x^*)$, $A(x^*)$ and $a(x^*)$, instead of those using $\bar{x}$, respectively. If x^* is 'sufficiently' close to $\bar{x}$, however, there may be a 'good' chance that x^* is feasible in (2.3). In $P(\bar{x};\gamma)$, we restrict the region of search by a hypercube around $\bar{x}$, where the parameter γ specifies the size of the cube. The reason for having the 'perturbation' vector ϵ in $P(\bar{x};\gamma)$ is purely technical; it is necessary to prove Theorems 2 and 3 below. In practice, we suggest that one set $\epsilon = 0$ (and also $\underline{\gamma} = 0$) to implement the procedure.

Let $\{x^i\}$ be the sequence of feasible solutions generated by the above procedure: an optimal solution $\hat{x}$ of $P(\bar{x};\gamma)$ will not be added to the sequence unless it is found to be feasible. The sequence is either finite or infinite. Clearly $c^{\mathrm{T}}x^i$ is monotone nonincreasing as i increases. One of the following three cases will occur.

Case 1: $\hat{x} = \bar{x}$ holds at some iteration (and the procedure terminates).

Case 2: The infinite sequence of x^i is generated.

Case 3: The optimization phase fails to generate a feasible $\hat{x}$ and the value of γ is indefinitely reduced. In this case the sequence $\{x^i\}$ is finite.

In the following, we shall prove that in Case 1 $\bar{x}$ is a local minimum, and that in Case 2 there exists a subsequence of $\{x^i\}$ converging a local minimum. Case 3 is a 'failure' case; we shall discuss a heuristic 'remedy' for it later on.

Theorem 2. *In Case* 1, $\bar{x}$ *is a local minimum of* (3.1).

Proof. Assume that $\epsilon > 0$ is fixed. By hypothesis, $\bar{x}$ solves $P(\bar{x};\gamma)$ for some $\gamma > 0$; i.e.,

$$c^{\mathrm{T}}\bar{x} = \min\left\{ c^{\mathrm{T}}x:\ \begin{array}{l} x \geq h, \quad -\gamma \leq x - \bar{x} \leq \gamma \\ \text{the solution } z \text{ of } (p(\bar{x}) + Rx,\, M(\bar{x})) \\ \text{satisfies } A(\bar{x})z \leq a(\bar{x}) + \epsilon \end{array} \right\}.$$

Let $\bar{z}$ denote the solution of the LCP $(p(\bar{x}) + R\bar{x}, M(\bar{x}))$. Suppose $\bar{x}$ is not a local minimum of (3.1). Then there exists a sequence $\{y^i\}$ of feasible solutions of (3.1) converging to $\bar{x}$ such that $c^{\mathrm{T}}y^i < c^{\mathrm{T}}\bar{x}$. For each i let z^i be the solution of the LCP $(p(\bar{x}) + Ry^i, M(\bar{x}))$. Since the solution of a LCP with a positive definite matrix is continuous in its data and since $y^i \to \bar{x}$, we have that z^i converges to $\bar{z}$. Since $\bar{x} \in X$, it holds that

$$A(\bar{x})\bar{z} \leq a(\bar{x}). \tag{3.2}$$

From $\epsilon > 0$ it follows that $A(\bar{x})z^i \leq a(\bar{x}) + \epsilon$ for a sufficiently large i. This means that y^i is feasible in $P(\bar{x};\gamma)$, contradicting the optimality of $\bar{x}$.

Theorem 3. *In Case 2, there exists a subsequence of* $\{x^i\}$ *converging to a local minimum of* (3.1).

Proof. Assume that $\epsilon > 0$ and $\underline{\gamma} > 0$ are fixed and $\gamma > \underline{\gamma}$ is kept throughout the procedure. Let $\{x^i\}$ be the sequence of feasible solutions generated during the procedure. Since $c^T x^i$ is nonnegative and decreasing, every x^i is contained in the compact region

$$X \cap \{x : c^T x \leq c^T x^1\}.$$

Thus, there exists a converging subsequence; let J denote the index set of one such subsequence. Let x^* denote the limit point of $\{x^j\}_{j \in J}$. Since X is closed, $x^* \in X$. Clearly, we have $c^T x^j \geq c^T x^*$ for each $j \in J$; indeed, $c^T x^i \geq c^T x^*$ for *every* $i \in \{1, 2, \ldots\}$.

Suppose x^* is not a local minimum. Then there exists a sequence $\{y^i\}$ of feasible solutions such that $\{y^i\}$ converges to x^* and $c^T y^i < c^T x^*$ for each $i \in \{1, 2, \ldots\}$. Since both $\{x^j\}_{j \in J}$ and $\{y^i\}$ converge to x^* it follows that $\|x^j - y^i\|$ converges to zero as i and $j \in J$ increase. For each $i \in \{1, 2, \ldots\}$ let z^i be the solution of the LCP $(p(y^i) + Ry^i, M(y^i))$; since $y^i \in X$,

$$A(y^i) z^i \leq a(y^i). \tag{3.3}$$

For every $i \in \{1, 2, \ldots\}$ and every $j \in J$, let z^{ij} denote the solution of the LCP $(p(x^j) + Ry^i, M(x^j))$.

Since p and M are continuous by assumption and since $\|x^j - y^i\|$ tends to zero, we have that both $\|p(x^j) - p(y^i)\|$ and $\|M(x^j) - M(y^i)\|$ converge to zero, as i and $j \in J$ increase. It then follows from the continuity of the solution of a LCP with a positive definite matrix in the problem data that $\|z^{ij} - z^i\|$ converges to zero as i and $j \in J$ increase. Further, the continuity of A and a implies that $\|A(x^j) - A(y^i)\|$ and $\|a(x^j) - a(y^i)\|$ both approach zero as i and $j \in J$ increase. This, coupled with (3.3), means that for sufficiently large i and $j \in J$ it holds that

$$A(x^j) z^{ij} \leq a(x^j) + \epsilon.$$

Therefore, if we choose i and $j \in J$ large enough, we have that y^i is feasible in the problem $P(x^j; \gamma)$, for any $\gamma > \underline{\gamma}$. But this contradicts the optimality of x^{j+1} because by construction $c^T x^{j+1} \geq c^T x^* > c^T y^i$.

During the execution of the iterative algorithm suppose the procedure 'stalls', or it fails to produce a feasible $\hat{x}$ solving $P(\bar{x}; \gamma)$ several iterations in a row. This 'stalling' is wasteful and may lead to the failure case (i.e., Case 3). When this happens we suggest one solve $P(\bar{x}; \gamma)$ using a negative vector ϵ. The idea is that an optimal solution to this modified problem has a better chance of belonging to X than that to $P(\bar{x}; \gamma)$ with a positive ϵ since the modified problem has more stringent constraints. In fact, if $-\epsilon$ is sufficiently large and if the optimal solution found for the modified problem is sufficiently close to $\bar{x}$, then the continuity of p,

M, A and a would ensure the feasibility. Of course, if $-\epsilon$ is too large, the modified problem may become infeasible; thus, an appropriate choice of ϵ is essential for this scheme to be successful. In the preliminary computational experiments we have conducted, this scheme was often found to be effective (see Appendix 2).

We have solved several small size numerical examples of Problem A using the iterative scheme. Here, we shall indicate the results on two of these problems (let us refer to them as P1 and P2 in the following). Both P1 and P2 arise from the 6-element truss problem sketched in Fig. 1; P1 and P2 differ from each other in the values of f_1 and f_2, and in the form of $a(x)$. In both problems $r = 2$, M is 6 by 6 and A is 4 by 6. We refer to [3] for a more complete description of the computational results of these and other problems.

We used a Fortran program called LPCPC for the Kaneko–Hallman algorithm (mentioned above) to perform each iteration of the procedure. For either P1 or P2 it took less than 0.2 second to execute each iteration. The procedure was terminated when a marked convergence was observed after 9 iterations for P1 and after 14 iterations for P2. Fig. 2 shows the transition of the objective value $c^T x^i$ for the sequence of feasible solutions $\{x^i\}$ generated by the procedure.

Sizes of the problems we solved were all very small; and so it is not possible to draw a conclusion as to how effective the proposed iterative procedure is for Problem A of practically meaningful sizes. We would like to point out, however, that among the problems we tried, we have a few 'failures' as well as 'successes' (such as P1 and P2 above). In a few cases, a 'stalling' of the procedure discussed above occurred which we could not resolve by using a negative ϵ. The procedure always performed successfully and well for those problems coming from well-posed structural engineering problems.

4. Problem B

The feasibility of Problem B can be determined easily by solving one linear program; thus we shall assume, hereafter, that Problem B is feasible. To analyze Problem B, (2.5), we shall first rewrite the problem using the optimal

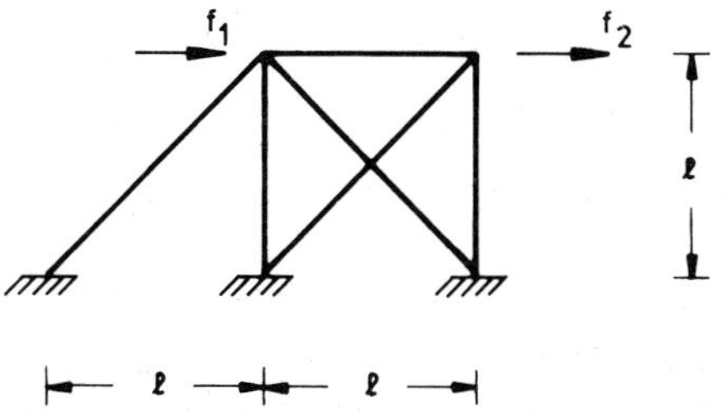

Fig. 1. 6-element truss.

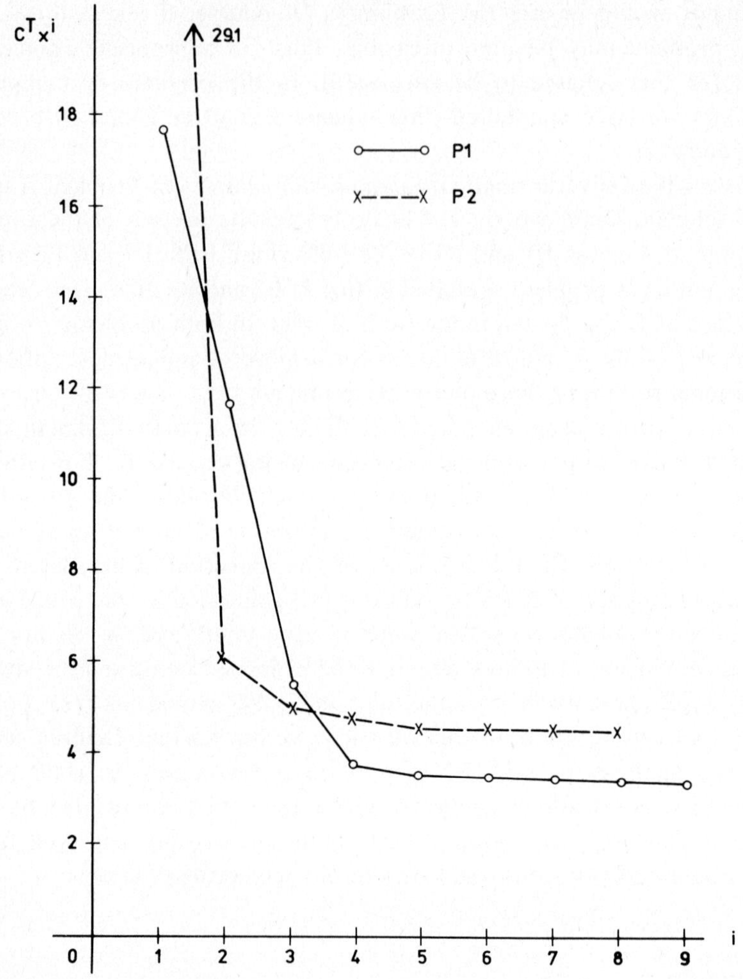

Fig. 2. Transition of the objection value.

values of a certain parametric QP. Let $\bar{\omega}$ and $\underline{\omega}$ be, respectively, the suprimum and minimum of the values of $\omega \geq 0$ such that (ω, z, y) satisfies (2.5b)–(2.5c) for some $z \geq 0$ and y. Clearly, $\underline{\omega}$ is finite (it may be zero), but $\bar{\omega}$ could be plus infinity. Let $\Omega = [\underline{\omega}, \bar{\omega}]$ if $\bar{\omega}$ is finite; let $\Omega = [\underline{\omega}, \infty)$ otherwise.

We consider the following parametric QP for $\omega \in \Omega$:

$$\min\{\tfrac{1}{2}z^{\mathrm{T}}Hz + K_0\colon Hz - N^{\mathrm{T}}y + k - \omega d \geq 0,\ C^{\mathrm{T}}y = f,\ z \geq 0\} \tag{4.1}$$

and let $r(\omega)$ be its optimal value for each fixed $\omega \in \Omega$. Note that $r(\omega)$ is well defined and finite since the QP is feasible for any given $\omega \in \Omega$ and its objective value is bounded from below. This parametric QP can be solved efficiently by some existing procedure, such as the parametric linear complementarity problem algorithm known as Cottle's algorithm [1]. For each $\omega \in \Omega$, let $(z(\omega), y(\omega))$ denote the optimal solution of the QP (4.1) computed by the algorithm. It is clear that $(z(\omega), y(\omega))$ is continuous and piecewise-linear.

For every fixed $\omega \in \Omega$, $\omega \neq 0$, let

$$s(\omega) = \min\left\{\frac{1}{\omega}(\tfrac{1}{2}z^{\mathrm{T}}Hz + K_0):\ Hz - N^{\mathrm{T}}y + k - \omega d \geq 0,\ C^{\mathrm{T}}y = f,\ z > 0\right\}.$$

Then, clearly, $s(\omega)$ equals $r(\omega)/\omega$ for every relevant ω and thus, Problem B may be repharsed as the one-dimensional problem:

$$\min\{s(\omega): \omega \in \Omega, \omega \neq 0\}. \tag{4.2}$$

If $\underline{\omega} > 0$ and $\bar{\omega} < \infty$, then an optimal solution exists since $r(\omega)$ is, and so $s(\omega)$ is, continuous in ω. The existence is not clear if $\underline{\omega} = 0$ and/or $\bar{\omega} = \infty$; so, we shall examine the behavior of $s(\omega)$ near zero and infinity.

First suppose $\underline{\omega} = 0$. In the neighborhood of $\omega = 0$, $z(\omega)$ is affine in ω; let

$$z(\omega) = b_0 + \omega c_0, \quad \omega > 0, \text{ small},$$

where b_0 and c_0 are constant n-vectors. Then, near zero, $s(\omega)$ is given by

$$s(\omega) = \alpha_0/\omega + \beta_0 + \delta_0\omega,$$

where $\alpha_0 = \frac{1}{2}b_0^{\mathrm{T}}Hb_0 + K_0 \geq 0$, $\beta_0 = b_0^{\mathrm{T}}Hc_0$ and $\delta_0 = \frac{1}{2}c_0^{\mathrm{T}}Hc_0 \geq 0$. Thus, there are two possibilities:

Case (a): $s(\omega) \nearrow \infty$ as $\omega \to 0+$ (if $\alpha_0 > 0$); and

Case (b): $s(\omega) \searrow \beta_0$ as $\omega \to 0+$ (if $\alpha_0 = 0$).

In either case, $s(\omega)$ approaches its limit monotonically. In Case (b), β_0 is nonnegative since $s(\omega)$ is nonnegative for all ω. If $s(\omega) > \beta_0$ for every ω, and if $\delta_0 > 0$ in Case (b), then we have that β_0 is the infimum of $s(\omega)$ on Ω ($\omega \neq 0$), but no ω attains the value. But in this case, we can choose $\omega > 0$ sufficiently small so that $s(\omega)$ is arbitrarily close to the infimum of $s(\omega)$; more specifically, for any given $\epsilon > 0$, choose ω_0 so that $\omega_0 \leq \epsilon/\delta_0$, then we have $s(\omega_0) \leq \inf_{\omega \in \Omega, \omega \neq 0} s(\omega) + \epsilon$.

Next, suppose $\bar{\omega} = \infty$. It is not difficult to see that there exists ω^* such that the optimal solution $(z(\omega), y(\omega))$ is affine in ω for all $\omega \geq \omega^*$ (if Cottle's algorithm is used, then ω^* is the last 'critical value'). For $\omega \geq \omega^*$, let $z(\omega) = b_1 + \omega c_1$, with constant n-vectors b_1 and c_1. Then, for $\omega \geq \omega^*$, we have

$$s(\omega) = \alpha_1/\omega + \beta_1 + \omega\delta_1,$$

where $\alpha_1 = \frac{1}{2}b_1^{\mathrm{T}}Hb_1 + K_0 \geq 0$, $\beta_1 = b_1^{\mathrm{T}}Hc_1$ and $\delta_1 = \frac{1}{2}c_1^{\mathrm{T}}Hc_1 \geq 0$. Two cases prevail:

Case (c): $s(\omega) \nearrow \infty$ as $\omega \to \infty$ (if $\delta_1 > 0$); and

Case (d): $s(\omega) \searrow \beta_1$ as $\omega \to \infty$ (if $\delta_1 = 0$).

Again, $s(\omega)$ is monotone and $\beta_1 \geq 0$ in Case (d). If $s(\omega) > \beta_1$ for all ω, and if $\delta_1 > 0$ in Case (d), then the infimum of $s(\omega)$ over Ω is β_1, but no finite ω attains it. In this case, we can choose ω sufficiently large so that $s(\omega)$ is arbitrarily close to the infimum value; more specifically, given any $\epsilon > 0$, choose ω_1 so that $\omega_1 > \alpha_1/\epsilon$ and then $s(\omega) \leq \inf_{\omega \in \Omega, \omega \neq 0} s(\omega) + \epsilon$.

Summarizing the results obtained in the above analysis we state:

Theorem 4. *Assume Problem* B *is feasible. An optimal solution to the problem always exists if $\bar{\omega}$ is finite and $\underline{\omega} > 0$. In the case the problem has no optimal solutions, we can choose ω so that the corresponding objective value $s(\omega)$ is arbitrarily close to the infimum of $s(\omega)$.*

Turning to methods to compute an optimal solution of Problem B, we shall indicate two approaches. It can be proved that the objective function is convex[3]: further, the function is differentiable and its derivatives are easy to evaluate. In addition, the constraints are all linear. Thus, the standard constrained nonlinear programming (NLP) algorithm should be effective, at least in principle. One potential difficulty is a possible numerical instability caused by the treatment of the constraint $\omega > 0$, when $\underline{\omega} = 0$. When one applies a standard NLP algorithm, one needs to choose a small positive ω' and replace $\omega > 0$ with $\omega \geq \omega'$. The obvious question is how small ω' should be? If it is too small, then it may lead to a numerical instability since the objective function contains the term $1/\omega$. We would take a risk of overlooking a true optimum if ω' is set too large. This trouble is particularly serious in the underlying engineering problem, since in that problem $\underline{\omega} = 0$ always holds and Case (a), above, is most likely.

We have solved a series of small size test problems of the form (2.5) arising from the engineering context. We used a computer program called GPM, which is a combination of gradient projection and quasi-Newton methods, available at the University of Wisconsin-Madison. Despite the good properties of the problem (i.e., convexity of the objective and the linearity of the constraints), and despite the smallness of the problem size (seven variables and six constrants), we ran into some numerical troubles, such as floating point overflows and violations of the upper limit of a quadratic approximation used in a line search subroutine inside the program, in many of the tested problems. We used $\omega' = 0.0001$. We applied the parametric QP approach, to be explained just below, to one of the troubled problems and the problem was solved rather easily, without any numerical difficulties.

An alternative approach would be to solve the line search problem (4.2), where the value of the objective function for any fixed ω can be evaluated by solving the QP (4.1). This approach is advantageous in at least two ways. First, in this approach, one never has to choose the artificial lower bound, ω', on ω, and thus would not have the potential numerical troubles present in the NLP approach. Also, the QP procedure is finite (see below), while most NLP algorithms are not. We note that the function $s(\omega)$ need not be convex on Ω, but is piecewise

[3] The author is indebted to R.W. Cottle for an interesting proof for this fact.

convex. We suggest the following procedure. Use Cottle's algorithm to solve the QP (4.1) 'completely', or for all value of $\omega \in \Omega$. Let $\{\omega^i\}$ be the sequence of critical values generated by the algorithm. Then, it is not difficult to show that $s(\omega)$ is convex and differentiable on $[\omega^i, \omega^{i+1}]$ for every i. Further, the minimum of $s(\omega)$ on each of these subintervals can be computed explicitly (easy to show it, by observing that $r(\omega)$ is quadratic on each subinterval). We have solved several test problems by this approach and obtained satisfactory results. These and other computational results (including those resulted in numerical trouble in the NLP approach) will be reported in a forthcoming paper by the author, Polizzotto and Mazzarella.

Appendix 1. A relationship between Problems A and B

The LCP (2.1) arises from the following set of relations which governs the behavior of elastic-plastic structures [4]:

$$C^{\mathrm{T}}y = f, \tag{A.1}$$

$$e + p = Cu, \tag{A.2}$$

$$y = Ke, \tag{A.3}$$

$$p = Nz, \tag{A.4}$$

$$w = Hz - N^{\mathrm{T}}y + k, \tag{A.5}$$

$$w \geq 0, \qquad z \geq 0, \qquad w^{\mathrm{T}}z = 0, \tag{A.6}$$

where the variables have the following meaning: y is the m-vector of *stresses*, e and p are, respectively, the m-vectors of *elastic* and *plastic strains*, z is the n-vector of *plastic intensities* and w is the n-vector of *plastic potentials*. From (A.1)–(A.4) it follows that

$$y = KC[C^{\mathrm{T}}KC]^{-1} + [KC[C^{\mathrm{T}}KC]^{-1}C^{\mathrm{T}}K - K]Nz;$$

and substituting y in (A.5) by this expression, the conditions (A.5)–(A.6) give rise to the LCP (q, M), where

$$q = -N^{\mathrm{T}}KC[C^{\mathrm{T}}KC]^{-1} + k \quad \text{and}$$
$$M = H + N^{\mathrm{T}}[K - KC[C^{\mathrm{T}}KC]^{-1}C^{\mathrm{T}}K]N.$$

In the rigid-plastic workhardening adaptation model for Problem B it is assumed that the elastic strains are negligible (compared to the plastic strains) and so we set $e = 0$ in (A.2) and neglect (A.3). The resulting conditions are:

$$C^{\mathrm{T}}y = f, \tag{A.7}$$

$$Nz = Cu, \tag{A.8}$$

$$w = Hz - N^{\mathrm{T}}y + k, \tag{A.9}$$

$$w \geq 0, \qquad z \geq 0, \qquad w^{\mathrm{T}}z = 0. \tag{A.10}$$

It will be shown that (A.7)–(A.10) are virtually the Kuhn–Tucker optimality conditions for Problem B.

To that end consider Problem B (2.5), except that the condition $\omega > 0$ is replaced with $\omega \geq 0$ (but with the understanding that ω is always positive at optimum or during the computation). We assume here that H is positive definite (as well as symmetric). The corresponding Kuhn–Tucker conditions can be written as

$$\frac{1}{\omega}\left(\frac{1}{2}z^{\mathrm{T}}Hz + K_0\right) = d^{\mathrm{T}}\xi \quad (\omega > 0), \tag{A.11}$$

$$H(z - \xi) \geq 0, \qquad z \geq 0, \qquad z^{\mathrm{T}}\mathrm{H}(z - \xi) = 0, \tag{A.12}$$

$$C^{\mathrm{T}}y = f, \tag{A.13}$$

$$N\xi = Cu, \tag{A.14}$$

$$w = Hz - N^{\mathrm{T}}y + k - \omega d, \tag{A.15}$$

$$w \geq 0, \qquad \xi \geq 0, \qquad w^{\mathrm{T}}\xi = 0, \tag{A.16}$$

where $(1/\omega)\xi$ and $(1/\omega)\mu$ are the Lagrange multipliers, respectively, for (2.5b) and (2.5c). Positive definiteness of H implies

$$(z - \xi)^{\mathrm{T}}H(z - \xi) \geq 0 \quad \text{or} \quad z^{\mathrm{T}}H(z - \xi) \geq \xi^{\mathrm{T}}H(z - \xi).$$

From (A.12) and $\xi \geq 0$ it follows that $H(z - \xi) = 0$, or $z = \xi$. Substituting ξ by z in (A.14) and (A.16) we see that (A.13)–(A.16) are the same as (A.7)–(A.10) except that $k - \omega d$ in (A.15) corresponds to k in (A.9). The equation (A.11) says that the best bound equals $d^{\mathrm{T}}z$. In fact, the vector d is chosen for the problem formulation is such a way that $d^{\mathrm{T}}z$ is the value of the deformation parameter and thus (A.11) says that the bound is tight.[4]

Appendix 2. Some examples

In this appendix, we shall present two examples of Problem A, where the use of 'negative ϵ' was effective in the iterative procedure. The first example was Problem A arising from a 3-element truss, where $r = 2$, M was 3 by 3 and A was 2 by 3. After producing two feasible $\hat{x}$ in the first two iterations, the optimization phase generated an infeasible $\hat{x}$ for three times in a row (the values of γ

[4] The relationship explained in this appendix has been explicitly stated and shown in papers by Polizzotto et al.; see [6] and also C. Polizzotto, "On workhardening adaptation of discrete structures under dynamic loadings", Archives of Mechanics 32 (1980) 81–99.

were $(0.3, 0.3)^T$, $(0.2, 0.2)^T$ and $(0.1, 0.1)^T$ during these three iterations, respectively, while $\bar{x} = (1.6, 0.6)^T$). The value of ϵ was always set equal to zero when this stalling was happening. We then used $\epsilon = (-0.01, -0.01)^T$ (and $\gamma = (0.1, 0.1)^T$) in the optimization phase which produced a feasible $\hat{x}$. This use of negative ϵ in fact saved a potentially serious stalling. To test the effect of ϵ we solved the same problem where ϵ was always set equal to zero; in this case 3 more consecutive infeasible $\hat{x}$ were generated where γ was reduced from $(0.1, 0.1)^T$ to $(0.05, 0.05)^T$ to $(0.02, 0.02)^T$ and to $(0.01, 0.01)^T$.

The other example was the same as P1 discussed above except for the value of the upper bound a. At some point during the procedure the optimal solution $\hat{x} = (0.905940,\ 1.353003)^T$, which was infeasible, was produced when $\bar{x} = (0.861831, 1.432211)^T$, $\gamma = (0.5, 1)^T$ and $\epsilon = 0$. In the next iteration we used the same γ (as well as the same $\bar{x}$) but $\epsilon = (-0.1, -0.1)^T$ and succeeded in producing a feasible $\hat{x}$. If we had decided to use $\epsilon = 0$ and had adopted the policy to halve the value of γ each time an infeasible $\hat{x}$ is produced, then we would have wasted *at least* 4 iterations without producing a feasible $\hat{x}$.

Acknowledgments

The author is grateful to one of the referees for pointing out that the minimum of $s(\omega)$ in Problem B on each subinterval can be obtained explicitly.

References

[1] R.W. Cottle, "Monotone solutions of the parametric linear complementarity problem," *Mathematical Programming* 3 (1972) 210–224.
[2] W.P. Hallman, "Complementarity in mathematical programming," Doctoral Dissertation, Department of Industrial Engineering, University of Wisconsin-Madison, WI (August 1979).
[3] I. Kaneko and G. Maier, "Optimum design of plastic structures under displacement constraints," *Computer Methods in Applied Mechanics and Engineering* 27 (1981) 369–391.
[4] G. Maier, "A matrix structural theory of piecewise linear elastoplasticity with interacting yield planes," *Meccanica* 5 (1970) 54–66.
[5] G. Maier, private comunication (July 1979).
[6] C. Polizzotto and C. Mazzarella, "Structural optimization based on workhardening adaptation concept," in: H.H.E. Leipholz, ed., *Structural control* (North-Holland, Amsterdam, 1980) pp. 597–612.

Mathematical Programming Study 17 (1982) 126–138.
North-Holland Publishing Company

ON THE CONVERGENCE OF A BLOCK SUCCESSIVE OVER-RELAXATION METHOD FOR A CLASS OF LINEAR COMPLEMENTARITY PROBLEMS

R.W. COTTLE[1]

Department of Operations Research, Stanford University, Stanford, CA 94305 U.S.A.

J.S. PANG[2]

Graduate School of Industrial Administration, Carnegie-Mellon University, Pittsburgh, PA 15213 U.S.A.

Received 8 August 1980
Revised manuscript received 1 June 1981

This paper develops a reduced block successive overrelaxation method for solving a class of (large-scale) linear complementarity problems. The main new feature of the method is that it contains certain reduction operations at each iteration. Such reductions are needed in order to ensure the boundedness (and therefore the existence of accumulation points) of the sequence of iterates produced by the algorithm. Convergence of the method is established by using a theorem due to Zangwill.

Key words: Convergence, Block Successive Overrelaxation Algorithms, Linear Complementarity Problem, Quadratic Programming, Compactness, Level Sets.

1. Introduction

The present research is motivated by an investigation (still in progress) of methods for solving a certain class of 'capacitated quadratic transportation problems'. One of these calls for the application of the block successive overrelaxation (BSOR) method [4] to the dual of the given problem. However, a technical problem is engendered by the unboundedness of the level sets of the dual objective function and the consequent breakdown of the convergence proof used in [4]. At issue is the existence of an accumulation point of the sequence of iterates produced by the algorithm. Fortunately, the structure of the problem permits a modification of the algorithm that leads to a remedy for this complication. Applying a simple transformation to the iterates forces the new points to lie in a compact set. Convergence of the algorithm can then be established by invoking a theorem of Zangwill [8].

[1] Research of this author and reproduction of this report were partially supported by the Department of Energy Contract DE-AC03-76SF00326, PA # DE-AT03-76ER72018 and the Office of the Naval Research Contract N00014-75-C-0267.

[2] Research of this author was supported by the Office of Naval Research under Contract N00014-75-c-0621 NR 047–048 and National Science Foundation Grant ECS-7926320.

Our purpose in this paper is to establish the convergence of the modified BSOR for a class of problems somewhat larger than that under consideration in the aforementioned study. To be precise, we concentrate our attention on a (large-scale) linear complementarity problem of the form: Find $y, v \in \mathbf{R}^N$ such that

$$v = f + FAc + FAF'y \geq 0, \qquad y \geq 0, \qquad v'y = 0. \tag{1}$$

The following blanket assumptions will be maintained throughout this paper:

(**A1**) The matrix $A \in \mathbf{R}^{p \times p}$ is symmetric and positive semi-definite.

(**A2**) There exists a vector x such that

$$FAx \leq f. \tag{2}$$

(**A3**) There exists an index set α such that for any y satisfying

$$AF'y = 0, \qquad f'y = 0, \qquad 0 \neq y \geq 0 \tag{3}$$

it follows that $y_j > 0$ if and only if $j \in \alpha$.

Remarks. (A3) holds vacuously if (3) has no solution. In fact, the nonexistence of a solution to (3) is equivalent to the so-called Slater condition, i.e., the consistency of the linear inequality system $FAx < f$.

Under Assumption (A1), the linear complementarity problem (1) is the set of Karush–Kuhn–Tucker conditions for the convex quadratic program

$$\begin{aligned} &\text{minimize} \quad \psi(y) = (f + FAc)'y + \tfrac{1}{2}y'FAF'y, \\ &\text{subject to} \quad y \geq 0. \end{aligned} \tag{4}$$

By using a basic property of convex quadratic programs (see [1], e.g.), it is easy to show that Assumptions (A1) and (A2) imply the existence of a solution to the problem (1), or equivalently, (4).

If the matrix A is in fact positive definite, then (4) is essentially the dual of the strictly convex quadratic program

$$\begin{aligned} &\text{minimize} \quad \phi(x) = c'x + \tfrac{1}{2}x'A^{-1}x, \\ &\text{subject to} \quad Fx \leq f. \end{aligned} \tag{5}$$

Note that (A2) implies the feasibility of (5) and thus the existence of an optimal solution.

Admittedly, the statement of Assumption (A3) is rather abstruse. It has a very intuitive geometrical interpretation, however. This says that the assumption holds if and only if the recession cone [7] of the nonempty level sets of the function ψ defined in (4) is a ray. We have chosen to state and use the assumption in the above algebraic format (instead of the geometrical one) because this is precisely how the assumption is shown satisfied in the capacitated quadratic transportation problem mentioned earlier. Indeed, one formulation of the problem leads to an LCP of the form (1) with A a positive diagonal matrix and

$$F = \begin{bmatrix} S \\ -D \\ I \\ -I \end{bmatrix} \tag{6}$$

where $\binom{S}{D}$ is the standard transportation constraint matrix and I is the identity matrix. The submatrix $\binom{I}{-I}$ arises from the capacity constraints on the flow variables. Obviously, the vector

$$\tilde{y} = \begin{bmatrix} E \\ e \\ 0 \\ 0 \end{bmatrix}$$

where e is the vector of ones, satisfies the system (3). By introducing a seemingly mild hypothesis on the capacities, one can show easily that (A3) holds. Indeed, the index set α can be identified trivially. It corresponds to the supply and demand constraints of the problem (cf. the vector $\tilde{y}$ given above).

In proving the convergence of iterative procedures for nonlinear programming, it is customary to require that the iterates lie in a compact set. The set in question is often a level set of the function being minimized. In the context of the quadratic program (4), the minimand is ψ. It is easy to see that the level sets of ψ are not bounded if the system (3) is consistent. In fact, if y^* is any vector satisfying $AF'y^* = 0$ and $f'y^* = 0$, then

$$\psi(y + \theta y^*) = \psi(y) \tag{7}$$

for all y and θ.

In the present paper, we shall show how the BSOR method described in [4] can be modified in such a way that the possible unboundedness of level sets will not affect the convergence of the method for solving (4)—or, equivalently, (1). Our analysis provides a unified treatment for both bounded and unbounded level sets. In particular, the analysis includes, as a special case, the recent study of Mangasarian [5] who treats the quadratic program (5) under a Slater condition.

2. On Assumption (A3)

Throughout the paper, we denote the linear complementarity problem (LCP)

$$w = q + Mz \geq 0, \quad z \geq 0 \quad \text{and} \quad w'z = 0$$

by the pair (q, M). In this section, we prove the geometrical interpretation to Assumption (A3) and derive a few related results.

To start, let C denote the set of all vectors y satisfying the system (3) and also containing the zero vector. Obviously, C is a (convex) cone. In fact, we have

Proposition 2.1. *The set C is the recession cone of all nonempty level sets $Y(\lambda) = \{y \geq 0 \colon \psi(y) \leq \lambda\}$ where $\psi(y)$ is the objective function in* (4).

Proof. Let $Y(\lambda)$ be a nonempty level set and let C' be its recession cone. Obviously $C \subseteq C'$ by (7). To prove the reverse inclusion, let $d \in C'$. Obviously, $d \geq 0$. Let $y \in Y(\lambda)$. Since the inequality

$$\psi(y + \theta d) = \psi(y + \theta d'(f + FAc + FAF'y) + \tfrac{1}{2}\theta^2 d'FAF'd \leq \lambda$$

must hold for all $\theta \geq 0$, we must have $d'FAF'd \leq 0$. By Assumption (A1), it follows that $AF'd = 0$. Hence $\psi(y + \theta d) \leq \lambda$ becomes $\psi(y) + \theta f'd \leq \lambda$. Since this must hold for all $\theta \geq 0$, we obtain $f'd \leq 0$. Assumption (A2) then implies that we must have $f'd = 0$. Hence $d \in C$ and therefore $C = C'$.

An immediate consequence of Proposition 2.1 is the following.

Corollary 2.2. *The set C is the recession cone of the solution set X of the problem* (1).

Proof. As noted in the last section, the problems (1) and (4) are equivalent. Thus, X, being the solution set of a convex program, is convex and is itself a level set of the form $Y(\lambda)$. It is also non-empty because (1)—and thus (4)—has a solution. The desired conclusion therefore follows from Proposition 2.1.

Corollary 2.3. *The LCP $(f + FAa, FAF')$ has a bounded solution set if and only if* (3) *is inconsistent.*

Proof. This follows immediately from Corollary 2.2.

Remark. Corollary 2.3 can also be proved by using an old result of Cottle [2] or a more recent characterization of Mangasarian [6].

In the next result, we establish the long-awaited geometrical interpretation of Assumption (A3); namely, the assumption holds if and only if the set C is a ray emerging from the origin.

Proposition 2.4. *Assumption* (A3) *holds if and only if there exists a nonnegative vector* y^* *such that* $C = \{y: y = \lambda y^* \text{ for some } \lambda \geq 0\}$.

Proof. Suppose C is of this form. If the system (3) is inconsistent, there is nothing to prove, so suppose it is consistent. This implies that the vector y^* must be nonzero. Let α be the set of indices which correspond to the nonzero components of y^* (i.e., its support). Obviously, if $y \in C \setminus \{0\}$, then $y_j > 0$ if and only if $j \in \alpha$.

Conversely, suppose that Assumption (A3) holds. If (3) is inconsistent it suffices to let y^* be the zero vector. On the other hand, if (3) is consistent, let y^* be any one of its solutions. Let $y \in C \setminus \{0\}$. Consider the vector $y - \lambda y^*$. For suitable $\lambda \geq 0$, the vector $y - \lambda y^*$ will belong to C and have at least one zero component, say the jth one with $j \in \alpha$. By (A3) this is impossible unless $y - \lambda y^*$ is the zero vector. This proves the proposition.

To describe the BSOR method, we let the rows of the matrix F be partitioned into blocks F_i $(i = 1, \dots, m)$. This induces a partitioning of $M = FAF'$ into submatrices $M_{ij} = F_i A F_j'$. Let the vector f be partitioned accordingly. Let J_i denote the set of indices of the rows in F_i (and f_i). Let n_i denote the cardinality of J_i, and finally (referring to (A3)) let

$$\alpha_i = \alpha \cap J_i, \quad i = 1, \dots, m.$$

Obviously, the following implication holds:

$$\left.\begin{aligned} AF_i' y_i &= 0 \\ f_i' y_i &= 0 \\ 0 \neq y_i &\geq 0 \end{aligned}\right\} \Rightarrow (y_i)_j > 0 \quad \text{if and only if } j \in \alpha_i.$$

The following result is a consequence of Assumption (A3).

Proposition 2.5. *Let assumption* (A3) *hold. Then for any partitioning of the rows of F, there can exist at most one index i for which the system*

$$AF_i' y_i = 0, \qquad f_i' y_i = 0, \quad 0 \neq y_i \geq 0 \tag{8$_i$}$$

is consistent.

Proof. Indeed, if there are indices $i_1 \neq i_2$ for which $(8)_{i_1}$ and $(8)_{i_2}$ are consistent, let $y^*_{i_1}$ and $y^*_{i_2}$ be solutions of these systems, respectively. Obviously, the vectors $y^1 = (y_l^1)$ and $y^2 = (y_l^2)$ with

$$y_l^1 = \begin{cases} 0, & \text{if } l \neq i_1, \\ y^*_{i_1}, & \text{if } l = i_1, \end{cases} \qquad y_l^2 = \begin{cases} 0, & \text{if } l \neq i_2, \\ y^*_{i_2}, & \text{if } l = i_2, \end{cases}$$

satisfy the system (3). By (A3), we must have $\alpha \subset J_{i_1} \cap J_{i_2} = \emptyset$. This contradiction establishes the proposition.

By Corollary 2.3, each subproblem $(f_i + F_iAa,\ F_iAF_i')$ has an unbounded solution set if and only if the corresponding system $(8)_i$ is consistent. It therefore follows from Proposition 2.5 that at most one of these m subproblems can have an unbounded solution set. (Incidentally, all the subproblems have nonempty solution sets.)

To close this section, we point out that in the capacitated quadratic transportation problem, the structure (6) of the matrix F induces a natural partitioning in which none of the systems $(8)_i$ is consistent. Thus, all the subproblems have bounded solution sets. In fact, the subproblems all have unique solutions.

3. Closedness of the component maps

The main tool used in our convergence result for the modified BSOR method is Convergence Theorem A of Zangwill [8]. To apply the theorem, it is necessary to show that the 'algorithmic map' involved is closed. In this section, we establish some preliminary results useful for this purpose.

The total number of rows in the matrix F is $N = \Sigma_1^m n_i$. For each $i = 1, \ldots, m$ let $y_i^* \in \mathbf{R}^{n_i}$ be *either* a fixed vector satisfying the system $(8)_i$ if the system is consistent *or* the zero vector if it is inconsistent. Similarly, let $y^* \in \mathbf{R}^N$ be either a fixed vector satisfying the system (3) if the system is consistent or the zero vector if it is not. Note that y^* may be different from $(y_1^*, \ldots, y_m^*)$. However, if $(y_1^*, \ldots, y_m^*)$ is nonzero, it can be used as y^*.

A vector $y_i \in \mathbf{R}_+^{n_i}$ is said to be *reduced* (with respect to the index set α_i) if at least one component in the subvector $(y_i)_{\alpha_i}$ is equal to zero. Then by $(8)_i$, y_i^* is either zero or *not* reduced. Let S_i denote the set of all reduced vectors in $\mathbf{R}_+^{n_i}$. The ith *reduction map* $\mathcal{R}_i : \mathbf{R}_+^{n_i} \to S_i$ is defined as follows:

$$\mathcal{R}_i(y_i) = y_i - \rho_i y_i^*$$

where

$$\rho_i = \rho_i(y_i) = \begin{cases} \min\{(y_i)_j/(y_i^*)_j : j \in \alpha_i\}, & \text{if } y_i^* \neq 0, \\ 0, & \text{otherwise.} \end{cases}$$

The reduction map $\mathcal{R}_i$ is well defined and continuous. If $y_i^* = 0$, then $\mathcal{R}_i$ is just the identity map. Similarly, by dropping the subscript i, we may define the reduced vectors in $\mathbf{R}_+^N$ as well as the complete reduction map $\mathcal{R} : \mathbf{R}_+^N \to S$, where S is the set of all reduced vectors in $\mathbf{R}_+^N$.

We define the ith *complementarity map* $\mathcal{C}_i : \mathbf{R}_+^N \to \mathbf{R}_+^N \times S_i$ as follows. Given $y = (y_1, \ldots, y_m) \in \mathbf{R}_+^N$, $\mathcal{C}_i(y)$ denotes the (nonempty) set of all points $(y, \mathcal{R}_i(\bar{y}_i))$ where $\bar{y}_i$ solves the LCP $(f_i + F_iAa,\ F_iAF_i')$ where

$$a = c + \sum_{l \neq i} F_l' y_l.$$

In general, $\mathcal{C}_i$ is a point-to-set map.

Note that if the subproblem has a unique solution, then by Corollary 2.3, $\mathcal{R}_i(\bar{y}_i) = \bar{y}_i$ so that the ith reduction is unnecessary. Roughly speaking, the motivation for including the reduction step in defining the map $\mathcal{C}_i$ is to ensure that $\mathcal{C}_i$ maps bounded sets into bounded sets.

Let $\omega^* < 2$ be a given positive scalar. Define the ith *relaxation map* $\mathcal{P}_i : \mathbf{R}_+^N \times \mathbf{R}_+^{n_i} \to \mathbf{R}_+^N$ as follows. For $(y, \hat{y}_i) \in \mathbf{R}_+^N \times \mathbf{R}_+^{n_i}$, the set $\mathcal{P}_i(y, \hat{y}_i)$ consists of all vectors of the form $(y_1, \ldots, y_{i-1}, \tilde{y}_i, y_{i+1}, \ldots, y_m)$ where $\tilde{y}_i = y_i + \tilde{\omega}(\hat{y}_i - y_i)$ for some $\tilde{\omega}$ such that

$$\min\{\omega^*, 1\} \le \tilde{\omega} \le \omega^* \quad \text{and} \quad \tilde{y}_i \ge 0.$$

The relaxation step in [4] is a particular realization of the relaxation map where $\tilde{\omega}$ is chosen as the largest possible value of ω for which

$$\omega \le \omega^* \quad \text{and} \quad y_i + \omega(\hat{y}_i - y_i) \ge 0.$$

A point to set map $\mathcal{M} : U \to V$ is *bounded* if for every subset $T \subset U$, the image $\bigcup \{\mathcal{M}(t): t \in T\}$ is a bounded subset of V.

Let $\mathcal{B}_i = \mathcal{P}_i \circ \mathcal{C}_i$ denote the composition of the ith complementarity and relaxation maps. In what follows, we show that $\mathcal{B}_i$ is a closed and bounded map from $\mathbf{R}_+^N$ into itself. We first prove this for $\mathcal{C}_i$.

For each index i and vector $a \in \mathbf{R}^p$ let $X_i(a)$ denote the set of all solutions of the LCP $(f_i + FAa, F_iAF_i')$.

Lemma 3.1. $\mathcal{R}_i(X_i(a)) = \{y_i \in X_i(a): \prod_{j\in\alpha_i}(y_i)_j = 0\}$.

Proof. For brevity, let T_i be the set on the right. Since $\rho_i(y_i) = 0$ for each $y_i \in T_i$, it follows that $T_i \subset \mathcal{R}_i(X_i(a))$. Conversely, let $\hat{y}_i = \mathcal{R}_i(\bar{y}_i)$ where $\bar{y}_i \in X_i(a)$. Then obviously, $\prod_{j\in\alpha_i}(\hat{y}_i)_j = 0$. It can easily be shown that $\hat{y}_i$ also solves the LCP $(f_i + FAa, F_iAF_i')$.

Proposition 3.2. *The ith complementarity map $\mathcal{C}_i$ is both closed and bounded.*

Proof. To show that $\mathcal{C}_i$ is closed let

$$y^k \to y, \quad y^k \in \mathbf{R}_+^N$$
$$z^k \to z = (y, \hat{y}_i), \quad z^k = (y^k, \mathcal{R}_i(\bar{y}_i^k)) \in \mathcal{C}_i(y^k).$$

As $\hat{y}_i$ is the limit of a sequence of reduced vectors $\hat{y}_i^k = \mathcal{R}_i(\bar{y}_i^k)$, it is itself reduced. It therefore suffices to prove that $\hat{y}_i \in X_i(a)$ where $a = c + \sum_{l\ne i} F_j' y_j$. Lemma 3.1 implies that for each k

$$\hat{y}_i^k \ge 0, \qquad \hat{v}_i^k = f_i + F_iA_a^k + F_iAF_i'y_i^k \ge 0, \qquad (\hat{v}_i^k)'(\hat{y}_i^k) = 0$$

where $a^k = c + \sum_{l\ne i} F_l' y_l^k$. Passing to the limit as $k \to \infty$, we obtain

$$\hat{y}_i \ge 0, \qquad \hat{v}_i = f_i + F_iAa + F_iAF_i'\hat{y}_i \ge 0, \qquad (\hat{v}_i)'(\hat{y}_i) = 0.$$

This establishes the closedness of $\mathscr{C}_i$. It is also bounded, for suppose the contrary. Then there exists a bounded subset $T \subset \mathbf{R}_+^N$ such that $\bigcup \{\mathscr{C}_i(t): t \in T\}$ is unbounded. Hence there exist sequences $\{y^k\} \in T$ and $\{z^k\} = (y^k, \mathscr{R}_i(\bar{y}_i^k))$ with $z^k \in \mathscr{C}_i(y^k)$ such that $\|z^k\| \to \infty$. Since $\{y^k\}$ is bounded, it has a convergent subsequence tending to some vector $y \in \mathbf{R}_+^N$. Without loss of generality, we may assume $y^k \to y$. Let $\hat{y}_i^k = \mathscr{R}_i(\bar{y}_i^k)$. Since $\|z^k\| \to \infty$ and $\{y^k\}$ is bounded, we must have $\|\hat{y}_i^k\| \to \infty$. However, the normalized sequence $\{\hat{y}_i^k / \|\hat{y}_i^k\|\}$ has a limit point, $\hat{y}_i$, and clearly $\hat{y}_i$ is reduced. Without loss of generality, we may assume $\hat{y}_i^k / \|\hat{y}_i^k\| \to \hat{y}_i$. For each k we have by Lemma 3.1

$$(\hat{y}_i^k)'(f_i + F_i A a^k + F_i A F_i' \hat{y}_i^k) = 0$$

where $a^k = c + \sum_{l \neq i} F_l' y_l^k$. Dividing the above equation by $\|\hat{y}_i^k\|^2$ and passing to the limit we obtain $(\hat{y}_i)' F_i A F_i' \hat{y}_i = 0$. By (A1) it follows that

$$AF_i'\hat{y}_i = 0. \tag{9}$$

Furthermore, we have $0 \geq (\hat{y}_i^k)'(f_i + F_i A a^k)$. Dividing by $\|\hat{y}_i^k\|$ and passing to the limit as $k \to \infty$, we obtain (in view of (9)) $(\hat{y}_i)' f_i \leq 0$. By Assumption (A2), we conclude that $\hat{y}_i$ satisfies

$$0 \neq \hat{y}_i \geq 0, \qquad (\hat{y}_i)' f_i = 0, \qquad AF_i'\hat{y}_i = 0.$$

Consequently, it follows that $(\hat{y}_i)_{\alpha_i} > 0$. But this contradicts the fact that $\hat{y}_i$ is reduced.

Proposition 3.3. *The ith relaxation map $\mathscr{P}_i$ is both closed and bounded.*

Proof. The boundedness is obvious. To show that $\mathscr{P}_i$ is closed, let $(y^k, \hat{y}_i^k) \to (y, \hat{y}_i)$ and $z^k \to z$ where

$$z^k = (y_1^k \ldots, y_{i-1}^k, \tilde{y}_i^k, y_{i+1}^k, \ldots, y_m^k) \in \mathscr{P}_i(y^k, \hat{y}_i^k)$$

and

$$z = (y_1, \ldots, y_{i-1}, \tilde{y}_i, y_{i+1}, \ldots, y_m).$$

It suffices to show that there exists a scalar $\tilde{\omega}$ with $\min\{\omega^*, 1\} \leq \tilde{\omega} \leq \omega^*$ such that $\tilde{y}_i = y_i + \tilde{\omega}(\hat{y}_i - y_i)$. But for each k, there exists a scalar $\tilde{\omega}^k \in [\min\{\omega^*, 1\}, \omega^*]$ such that $\tilde{y}_i^k = y_i^k + \tilde{\omega}^k(\hat{y}_i^k - y_i^k) \geq 0$. Since the $\tilde{\omega}^k$ lie in a compact interval they have a limit point $\tilde{\omega}$. This $\tilde{\omega}$ will do.

Lemma 3.4. *The composition of two bounded (point-to-set) maps is bounded.*

Proof. Indeed, if $\mathscr{M}_1 : U \to V$ and $\mathscr{M}_2 : V \to W$ are two bounded (point-to-set) maps and T is a bounded subset of U, then the set

$$\mathscr{M}_2 \circ \mathscr{M}_1(T) = \bigcup \{\mathscr{M}_2(s): s \in \mathscr{M}_1(T)\}$$

is obviously bounded.

Combining these results, we obtain immediately

Proposition 3.5. *The map $\mathcal{B}_i$ is both closed and bounded.*

Proof. This follows from Propositions 3.2 and 3.3 by applying Lemma 3.4 and [8, Lemma 4.2].

Remark. The boundedness of the complementarity map $\mathcal{C}_i$ is crucial in order to apply [8, Lemma 4.2] to deduce that $\mathcal{B}_i$ is closed. For the same reason, the boundedness of $\mathcal{B}_i$ is important in proving the closedness of the algorithmic map to be given later. The role played by the reduction maps $\mathcal{R}_i$ in these deductions should now be very transparent.

We point out that a vector $z \in \mathcal{B}_i(y)$ might not be reduced with respect to α_i. This is because the relaxation map $\mathcal{P}_i$ does not necessarily preserve 'reducedness'.

4. The reduced BSOR algorithm

In its simplest form, the modified version of the BSOR method for solving the LCP (1) can be described by its associated algorithmic map

$$\mathcal{A} = \mathcal{R} \circ \mathcal{B}_m \circ \ldots \circ \mathcal{B}_1. \tag{10}$$

More precisely, given an arbitrary non-negative vector $\hat{y}^0$, the algorithm generates a sequence $\{\hat{y}^k\}$ of vectors as follows. If $\hat{y}^k$ solves the problem (1), stop; otherwise pick a vector $\hat{y}^{k+1} \in \mathcal{A}(\hat{y}^k)$ and repeat. For an obvious reason, we call this the reduced BSOR algorithm. It is clear that any fixed point of the map $\mathcal{A}$ solves the LCP (1). (See Corollary 4.3.)

There are essentially two new features in this reduced BSOR algorithm. First, a (possibly unnecessary) reduction is performed after each linear complementarity subproblem is solved. (The precise manner in which these subproblems are solved is optional.) Second, at the end of each iteration, a complete reduction (defined by the map $\mathcal{R}$) is performed. We have seen how reductions of the first kind are useful. Basically, the second kind of reduction is needed for a similar reason; namely, to ensure the boundedness of the sequence $\{\hat{y}^k\}$ generated by the algorithm.

Our principal convergence result for the reduced BSOR algorithm is

Theorem 4.1. *Applied to the LCP* (1) *for which* (A1), (A2) *and* (A3) *are satisfied, the reduced BSOR algorithm either terminates with a solution or else the sequence of iterates contains an accumulation point which solves the problem.*

We first establish three preliminary results. The first extends [4, Theorem 1]

Lemma 4.2. *Let*

$$\phi(x_1, x_2) = \tfrac{1}{2}\begin{pmatrix} x_1 \\ x_2 \end{pmatrix}'\begin{pmatrix} M_{11} & M_{12} \\ M_{21} & M_{22} \end{pmatrix}\begin{pmatrix} x_1 \\ x_2 \end{pmatrix} + \begin{pmatrix} r \\ s \end{pmatrix}'\begin{pmatrix} x_1 \\ x_2 \end{pmatrix}$$

where M_{11}, $M_{12} = M'_{21}$, M_{22}, r and s are given, M_{11} is symmetric and positive semi-definite, and x_1, x_2 are vector variables. Let $\bar{x}_1$ solve the LCP $(r + M_{12}\bar{x}_2, M_{11})$ for some vector $\bar{x}_2$. Then for all $x_1 \geq 0$ and all $\omega \in (0, 2)$

$$\phi(x_1 + \omega(\bar{x}_1 - x_1), \bar{x}_2) \leq \phi(x_1, \bar{x}_2)$$

with equality if and only if x_1 also solves the LCP $(r + M_{12}\bar{x}_2, M_{11})$.

Proof. Let $\delta = (\phi(x_1 + \omega(\bar{x}_1 - x_1), \bar{x}_2) - \phi(x_1, \bar{x}_2))/\omega$. By an easy manipulation, we obtain

$$\begin{aligned}\delta &= (\bar{x}_1 - x_1)'(r + M_{12}\bar{x}_2) + (\bar{x}_1 - x_1)'M_{11}x_1 + \tfrac{1}{2}\omega(\bar{x}_1 - x_1)'M_{11}(\bar{x}_1 - x_1) \\ &\leq (\bar{x}_1 - x_1)'(r + M_{12}\bar{x}_2) + (\bar{x}_1 - x_1)'M_{11}x_1 + (\bar{x}_1 - x_1)'M_{11}(\bar{x}_1 - x_1) \\ &= \bar{x}'_1(r + M_{11}\bar{x}_1 + M_{12}\bar{x}_2) - x'_1(r + M_{11}\bar{x}_1 + M_{12}\bar{x}_2) \leq 0.\end{aligned}$$

If $\phi = 0$, then

$$(\bar{x}_1 - x_1)'M_{11}(\bar{x}_1 - x_1) = x'_1(r + M_{11}\bar{x}_1 + M_{12}\bar{x}_2) = 0$$

Since M_{11} is symmetric and positive semi-definite, $M_{11}x_1 = M_{11}\bar{x}_1$ and hence, x_1 solves the linear complementarity problem $(r + M_{12}\bar{x}_2, M_{11})$ as well.

Corollary 4.3. *For any $y \in \mathbf{R}^N_+$ and $z \in \mathcal{A}(y)$, $\psi(z) \leq \psi(y)$ with equality if and only if y solves (1).*

Proof. This follows easily from the definition of $\mathcal{A}$ and repeated use of Lemmas 3.1 and 4.2.

Lemma 4.4. *The sequence $\{\hat{y}^k\}$ of iterates generated by the reduced BSOR algorithm is bounded.*

Proof. By Corollary 4.3, we have for each k $\psi(\hat{y}^k) \leq \lambda = \psi(\hat{y}^0)$. The remainder of the proof resembles that of Proposition 3.2 and is omitted.

Proof of Theorem 4.1. By repeated use of Proposition 3.5 and [8, Lemma 4.2] one can easily show that the algorithmic map is closed. The desired conclusion now follows from Lemma 4.4, Corollary 4.3 and [8, Convergence Theorem A].

5. An extension

It is rather easy to extend the reduced BSOR algorithm to treat the following generalization of the quadratic program (4): Find a vector $y \in \mathbf{R}^N$ to

$$\text{minimize} \quad \psi(y) = q'y + \tfrac{1}{2}y'My,$$
$$\text{subject to} \quad y_i \in Y_i, \quad i = 1, \dots, m. \tag{11}$$

Here the vector y is partitioned into subvectors $y_i \in \mathbf{R}^{n_i}$ and each Y_i is a nonempty polyhedral set in $\mathbf{R}^{n_i}$

$$Y_i = \{y_i \in \mathbf{R}^{n_i}: B_i y_i \le b_i\}$$

where B_i and b_i are arbitrary matrices and vectors respectively. The matrix M in (11) is symmetric and positive semi-definite and is partitioned into submatrices M_{ij} $(i, j = 1, \dots, m)$ where each M_{ij} is n_i by n_j. The vector q is partitioned accordingly.

Without repeating many fo the details, we shall in what follows simply present the generalized version of (A3), define the component maps and state the main theorem of convergence for the algorithm. We point out that the program (11) includes as a special case the one treated in [3]. In the latter program, each Y_i is a closed interval of $\mathbf{R}$ and the matrix M is symmetric and positive definite.

For $i = 1, \dots, m$, let J_i denote the set of indices in the subvector y_i and let 0^+Y_i denote the recession cone of the set Y_i, i.e.,

$$0^+Y_i = \{d_i \in \mathbf{R}^{n_i}: B_i d_i \le 0\}.$$

Define

$$C_i = \{d_i \in \mathbf{R}^{n_i}: q_i' d_i = 0,\ M_{ii} d_i = 0\} \cap 0^+Y_i$$

and let

$$C = \{d \in \mathbf{R}^N: q'd = 0,\ Md = 0\} \cap \prod_{i=1}^{m} 0^+ y_i.$$

Let B denote the block diagonal matrix whose diagonal blocks are the B_i. Finally, let $Y = \prod_{i=1}^m Y_i$ be the feasible set of the program (11). We state the generalized version of Assumption (A3):

(A4) There exists a nonempty index set α such that for any vector $d \in C \setminus \{0\}$, it follows that $(Bd)_j < 0$ if and only if $j \in \alpha$.

For each $i = 1, \dots, m$, let $\alpha_i = \alpha \cap J_i$. Let d_i^* be a vector in $C_i \setminus \{0\}$ if $C_i \ne \{0\}$ or the zero vector if $C_i = \{0\}$. Similarly, let d^* be a vector in $C \setminus \{0\}$ if $C \ne \{0\}$ or the zero vector if $C = \{0\}$.

As in the previous case, it can be shown easily that if the program (11) has an optimal solution, then the set C is the recession cone of all nonempty level sets

$$Y(\lambda) = \{y \in Y: \psi(y) \le \lambda\}.$$

Moreover, Assumption (A4) holds if and only if either $C = \{0\}$ or there exists a vector d^* with $Bd^* \ne 0$ such that

$$C = \{d \in \mathbf{R}^N : d = \lambda d^* \text{ for some } \lambda \geq 0\}$$

i.e., the cone C is either $\{0\}$ or a ray.

A vector $y_i \in Y_i$ is said to be *reduced* (with respect to the index set α_i) if at least one component in the subvector $(b_i - B_i y_i)_{\alpha_i}$ is zero. Let S_i denote the set of reduced vectors in Y_i. The *reduction map* $\mathcal{R}_i : Y_i \to S_i$ is defined as follows:

$$\mathcal{R}_i(y_i) = y_i - \rho_i d_i^*$$

where

$$\rho_i = \rho_i(y_i) = \begin{cases} \min\{(B_i y_i - b_i)_j / (B_i d_i^*)_j : j \in \alpha_i\}, & \text{if } d_i^* \neq 0, \\ 0, & \text{otherwise.} \end{cases}$$

Similarly, by dropping the subscript i, we may define reduced vectors in Y and the complete reduction map $\mathcal{R} : Y \to S$, where S is the set of reduced vectors in Y.

Extending the ith complementarity map, we define the ith *subprogram map* $\mathcal{S}_i : Y \to Y \times S_i$. Given $y = (y_i, \ldots, y_m) \in Y$, $\mathcal{S}_i(y)$ denotes the set of all points $(y, \mathcal{R}_i(\bar{y}))$ where $\bar{y}_i$ solves the quadratic program

$$\begin{aligned} &\text{minimize} \quad \Big(q_i + \sum_{j \neq i} M_{ij} y_j\Big)' z_i + \tfrac{1}{2} z_i' M_{ii} z_i, \\ &\text{subject to} \quad z_i \in Y_i. \end{aligned} \tag{12$_i$}$$

It is important to note that the set Y_i is included as the feasible region of the ith subprogram.

Finally, the ith *relaxation map* $\mathcal{P}_i : Y \times Y_i \to Y$ is defined as follows. Let $\omega^* < 2$ be a given positive scalar. For $(y, \bar{y}_i) \in Y \times Y_i$, the set $\mathcal{P}_i(y, \bar{y}_i)$ consists of all vectors of the form

$$(y_1, \ldots, y_{i-1}, \tilde{y}_i, y_{i+1}, \ldots, y_m)$$

where $\tilde{y}_i = y_i + \tilde{\omega}(\bar{y}_i - y_i)$ for some $\tilde{\omega}$ such that $\min\{\omega^*, 1\} \leq \tilde{\omega} \leq \omega^*$ and $\tilde{y}_i \in Y_i$.

Let $\mathcal{B}_i = \mathcal{P}_i \circ \mathcal{S}_i$ be the ith component map. The algorithmic map $\mathcal{A}$ is defined by (10). The main convergence theorem is the following.

Theorem 5.1. *Suppose that the quadratic program* (11) *has an optimal solution and that Assumption* (A4) *holds. Then, provided that the initial vector* $\hat{y}^0$ *is feasible, the same conclusion of Theorem* 4.1 *holds for the reduced BSOR algorithm applied to the program* (11).

Remark. The assumption that the program (11) has an optimal solution is not crucial for the applicability of the algorithm. In fact, without the assumption, the algorithm can still be applied but may terminate at a situation where a certain subprogram (12)$_i$ has an unbounded objective function value. It is easy to show that if this happens, then the original program (11) must have an unbounded objective as well.

6. Concluding remarks

This paper is intended to provide the theoretical foundations for the reduced BSOR method which is one of the algorithms being considered in our investigation of computational procedures for solving the capacitated quadratic transportation problem. Preliminary computational experience with problems of considerable size (e.g., $N \approx 5000$) suggests that the reduced BSOR method may prove efficient in this application and possibly others as well. We plan to report our computational results elsewhere.

References

[1] R.W. Cottle, "Note on a fundamental theorem in quadratic programming", *Journal of the Society for Industrial and Applied Mathematics* 12 (1964) 663–665.

[2] R.W. Cottle, "Solution rays for a class of complementarity problems", *Mathematical Programming Study* 1 (1974) 59–70.

[3] R.W. Cottle and M.S. Goheen, "A special class of large quadratic programs", in: O.L. Mangasarian, R.R. Meyer and S.M. Robinson, eds., *Nonlinear programming* 3 (Academic Press, New York, 1978) pp. 361–389.

[4] R.W. Cottle, G.H. Golub and R.S. Sacher, "On the solution of large, structured linear complementarity problems: The block partitioned case", *Applied Mathematics and Optimization* 4 (1978) 347–363.

[5] O.L. Mangasarian, "Iterative solution of linear programs", Technical Report #327, Computer Sciences Department, University of Wisconsin, Madison, WI (June 1979).

[6] O.L. Mangasarian, "Characterization of bounded solutions of linear complementarity problems", Technical Report #359, Computer Sciences Department, University of Wisconsin, Madison, WI (December 1979).

[7] R.T. Rockafellar, *Convex analysis* (Princeton University Press, Princeton, 1970).

[8] W.I. Zangwill, *Nonlinear programming: A unified approach* (Prentice-Hall, Englewood Cliffs, NJ, 1969).

Mathematical Programming Study 17 (1982) 139–159.
North-Holland Publlishing Company

PROJECTION METHODS FOR VARIATIONAL INEQUALITIES WITH APPLICATION TO THE TRAFFIC ASSIGNMENT PROBLEM*

Dimitri P. BERTSEKAS and Eli M. GAFNI

Laboratory for Information and Decision Systems, Massachussetts Institute of Technology, Cambridge, MA 02139, U.S.A.

Received 1 August 1980

It is well known [2, 3, 16] that if $\bar{T}:\mathbf{R}^n \to \mathbf{R}^n$ is a Lipschitz continuous, strongly monotone operator and X is a closed convex set, then a solution $x^* \in X$ of the variational inequality $(x - x^*)'\bar{T}(x^*) \geq 0$, $\forall\, x \in X$ can be found iteratively by means of the projection method $x_{k+1} = P_X[x_k - \alpha \bar{T}(x_k)]$, $x_0 \in X$, provided the stepsize α is sufficiently small. We show that the same is true if $\bar{T}$ is of the form $\bar{T} = A'TA$, where $A:\mathbf{R}^n \to \mathbf{R}^m$ is a linear mapping, provided $T:\mathbf{R}^m \to \mathbf{R}^m$ is Lipschitz continuous and strongly monotone, and the set X is polyhedral. This fact is used to construct an effective algorithm for finding a network flow which satisfies given demand constraints and is positive only on paths of minimum marginal delay or travel time.

Key words: Projection Methods, Variational Inequalities, Traffic Assignment, Network Routing, Multicommodity Network Flows.

1. Introduction

We consider the problem of finding $x^* \in X$ satisfying the variational inequality

$$(x - x^*)'A'T(Ax^*) \geq 0, \quad \forall\, x \in X \tag{1}$$

where X is a nonempty subset of $\mathbf{R}^n$, A is a given $m \times n$ matrix and $T:\mathbf{R}^m \to \mathbf{R}^m$ is a nonlinear operator which is Lipschitz continuous and strongly monotone in the sense that there exist positive scalars L and λ such that for all y_1, y_2 in the set $Y = \{y \mid y = Ax, x \in X\}$ we have

$$|T(y_1) - T(y_2)| \leq L|y_1 - y_2|, \tag{2}$$

$$[T(y_1) - T(y_2)]'(y_1 - y_2) \geq \lambda |y_1 - y_2|^2. \tag{3}$$

In the relations above and throughout the paper all vectors are considered to be column vectors, and a prime denotes transposition. The standard norm in $\mathbf{R}^n$ is denoted $|\cdot|$, i.e., $|x| = (x'x)^{1/2}$ for all $x \in R^n$.

We are interested in the projection algorithm

$$x_{k+1} = P_X^S[x_k - \alpha S^{-1}A'T(Ax_k)], \quad x_0 \in X \tag{4}$$

* Work supported by Grants ONR–N00014-75-C-1183 and NSF ENG-7906332.

where $\alpha > 0$ is a stepsize parameter, S is a fixed positive definite symmetric matrix, and, for any $z \in \mathbf{R}^n$, $P_X^S(z)$ denotes the unique projection of z on the set X with respect to the norm $\|\cdot\|$ corresponding to S

$$\|w\| = (w'Sw)^{1/2}, \quad \forall\, w \in R^n. \tag{5}$$

The variational inequality (1) arises from the variational inequality

$$(y - y^*)'T(y^*) \geq 0, \quad \forall\, y \in Y \tag{6}$$

through the transformation

$$y = Ax, \qquad Y = AX = \{y \mid y = Ax, x \in X\}. \tag{7}$$

It is possible to employ the projection algorithm

$$y_{k+1} = P_Y^Q[y_k - \alpha Q^{-1}T(y_k)], \quad y_0 \in Y \tag{8}$$

for solving (6) where α is a positive stepsize parameter and Q is positive definite symmetric. It has been shown by Sibony [16] that if T is Lipshitz continuous and strongly monotone, and α is chosen sufficiently small, then the sequence $\{y_k\}$ generated by (8) converges to the unique solution y^* of (6). The rate of convergence is typically linear although a superlinear convergence rate is possible in exceptional cases [9]. Strong monotonicity of the mapping T is an essential assumption for these results to hold. Our motivation for considering iteration (4) stems from the fact that in some cases projection on the set Y is very difficult computationally while projection on the set X through the transformation $y = Ax$ [cf. (7)] may be very easy. Under these circumstances if all other factors are equal, the projection method (4) is much more efficient than the method (8). This situation occurs for example in the application discussed in Section 3.

A potential difficulty with the transformation idea described above is that the mapping $A'TA$ is not strongly monotone unless the matrix $A'A$ is invertible. Thus convergence of iteration (4) is not guaranteed by the existing theory [2, 3, 16]. One of the contributions of this paper is to show that the convergence and rate of convergence properties of iteration (4) are satisfactory and comparable with those of iteration (8). These results hinge on the assumption that X is polyhedral, and it is unclear whether and in what form they hold if X is a general convex set.

In Section 3 we consider a classical traffic equilibrium problem arising in several contexts including communication and transportation networks, which can be modelled in terms of a variational inequality of the form (1). A projection algorithm for solving this problem which is essentially of the form (8) has been given by Dafermos [8]. Her algorithm however operates in the space of link flows, and involves a projection iteration which is very costly for large networks. We consider an alternative algorithm which is basically of the form (4) and operates in the space of path flows. Because the projection iteration can be carried out easily in this space our algorithm is much more efficient. An algorithm which has several similarities with ours has been proposed by Aashtiani [1] and has performed well in computational experiments. However, Aashtiani's algorithm cannot be shown to converge in general. By contrast the results of Section 2 guarantee convergence and linear rate of convergence for our method. There are also other methods [4–7, 11–13] for solving the special case of the traffic assignment problem where T is a gradient mapping and there is an underlying convex programming problem. Some of these methods [4–6, 12, 13] are of the projection type. The algorithm of the present paper, however, seems to be the first that can solve the general problem, is suitable for large networks, and is demonstrably convergent.

In Section 4 we consider the generalized version of iteration (4),

$$x_{k+1} = P_X^{S_k}[x_k - \alpha S_k^{-1} A'T(Ax_k)], \quad x_0 \in X$$

where $\{S_k\}$ is a sequence of positive definite symmetric matrices with eigenvalues bounded above and bounded away for zero. Projection algorithms of this type include Newton's method for constrained minimization [10, 14], and several network flow algorithms [1, 4, 5, 13]. Except for the case where T is the gradient of a convex function, there are no convergence results in the literature for this algorithm, even when A is the identity matrix. We show that if care is taken in the way the matrices S_k are allowed to change, then the resulting algorithm is convergent at a rate which is at least linear. We also provide a computational example involving a traffic assignment problem.

2. Projection methods for variational inequalities

Let X^* be the set of all solutions of the variational inequality (1). We have that X^* is polyhedral and is given by

$$X^* = \{x \in X \mid Ax = y^*\} \tag{9}$$

where y^* is the unique solution of the variational inequality (6). For any $x \in \mathbf{R}^n$ we denote by $p(x)$ its unique projection on X^* with respect to the norm (5), i.e.,

$$p(x) = \arg\min\{\|x - z\| \mid z \in X^*\}. \tag{10}$$

We recall that projection on a convex set is a nonexpansive mapping (see e.g. [9]), so we have

$$\|p(x_1) - p(x_2)\| \le \|x_1 - x_2\|, \quad \forall\, x_1, x_2 \in \mathbf{R}^n. \tag{11}$$

We denote the mapping $A'TA$ by $\bar{T}$, i.e.,

$$\bar{T}(x) = A'T(Ax), \quad \forall\, x \in \mathbf{R}^n. \tag{12}$$

With this notation we have

$$(x - x^*)'\bar{T}(x^*) \geq 0, \quad \forall\, x \in X,\, x^* \in X^*. \tag{13}$$

We also denote for all $x \in X$, $\alpha > 0$

$$\hat{x}(x, \alpha) = P_X^S[x - \alpha S^{-1}\bar{T}(x)]. \tag{14}$$

With this notation the projection iteration (4) is written as

$$x_{k+1} = \hat{x}(x_k, \alpha), \quad x_0 \in X. \tag{15}$$

It can be seen that we can also obtain x_{k+1} as the unique solution to the problem

$$\begin{aligned} &\text{minimize} \quad (x - x_k)'\bar{T}(x_k) + \frac{1}{2\alpha}(x - x_k)'S(x - x_k), \\ &\text{subject to} \quad x \in X. \end{aligned} \tag{16}$$

In view of (13) it is easy to see that for all solutions $x^* \in X^*$ and $\alpha > 0$ we have

$$\hat{x}(x^*, \alpha) = x^* = p(x^*). \tag{17}$$

Thus if $x_k \in X^*$ for some k the algorithm (15) essentially terminates.

Our main result is given in the following proposition.

Proposition 1. *Assume that T is Lipschitz continuous and strongly monotone, and X is polyhedral.*

(a) *There exist positive scalars $q(S)$ and $r(S)$ depending continuously on S, such that for all $x \in X$ and $\alpha > 0$*

$$\|\hat{x}(x, \alpha) - p[\hat{x}(x, \alpha)]\|^2 \leq [1 - 2\alpha q(S) + \alpha^2 r(S)]\|x - p(x)\|^2. \tag{18}$$

(b) *There exists $\bar{\alpha} > 0$ such that for all $\alpha \in (0, \bar{\alpha}]$ the sequence $\{x_k\}$ generated by iteration (15) converges to a solution x^* of the variational inequality (1). The rate of convergence is at least linear in the sense that for each $\alpha \in (0, \bar{\alpha}]$ and $x_0 \in X$ there exist scalars β (depending on α) and q (depending on α and x_0) such that $q > 0$, $\beta \in (0, 1)$, and*

$$\|x_k - x^*\| \leq q\beta^k, \quad k = 0, 1, \ldots$$

The proof of Proposition 1 relies on the following lemma, the proof of which is relegated to the appendix. The lemma is easy to conjecture in terms of geometrical arguments (see Fig. 1).

Lemma 1. *Assume the conditions of Proposition 1 hold. Let*

$$\tilde{X}^* = \{x \in \mathbf{R}^n \mid Ax = y^*\}, \tag{19}$$

$$\tilde{p}(x) = \arg\min\{\|x - z\| \mid z \in \tilde{X}^*\}. \tag{20}$$

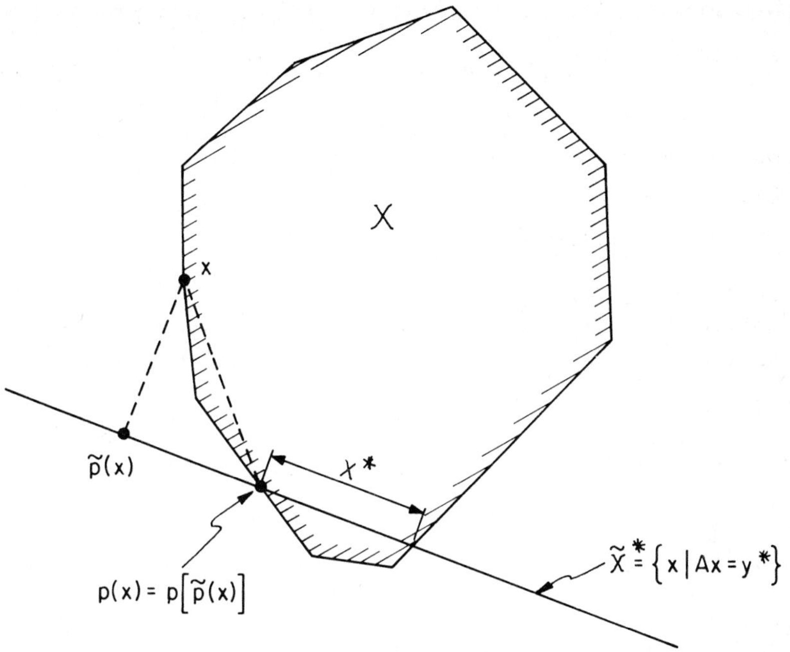

Fig. 1. Geometric interpretation of Lemma 1.

Then $p(x) = p[\tilde{p}(x)]$ *and then there exists a positive scalar* $\eta(S)$ *depending continuously on S such that for all* $x \in X$

$$\|x - \tilde{p}(x)\|^2 \geq \eta(s)\|p(x) - \tilde{p}(x)\|^2. \tag{21}$$

Proof of Proposition 1. (a) From (10) we obtain

$$\|\hat{x}(x, \alpha) - p[\hat{x}(x, \alpha)]\|^2 \leq \|\hat{x}(x, \alpha) - p(x)\|^2. \tag{22}$$

Also, using (17) and (11) we have

$$\begin{aligned}\|\hat{x}(x, \alpha) - p(x)\|^2 &= \|\hat{x}(x, \alpha) - \hat{x}[p(x), \alpha]\|^2 \\ &\leq \|[x - p(x)] - \alpha S^{-1}[\bar{T}(x) - \bar{T}[p(x)]]\|^2 \\ &= \|x - p(x)\|^2 - 2\alpha[\bar{T}(x) - \bar{T}[p(x)]]'[x - p(x)] \\ &\quad + \alpha^2\|S^{-1}[\bar{T}(x) - \bar{T}[p(x)]]\|^2.\end{aligned} \tag{23}$$

In view of (22) and (23) it will be sufficient to show that there exist positive scalars $q(S)$ and $r(S)$ depending continuously on S such that

$$[\bar{T}(x) - \bar{T}[p(x)]]'[x - p(x)] \geq q(S)\|x - p(x)\|^2, \quad \forall x \in X, \tag{24a}$$

$$\|S^{-1}[\bar{T}(x) - \bar{T}[p(x)]]\|^2 \leq r(S)\|x - p(x)\|^2, \quad \forall x \in X. \tag{24b}$$

To prove (24a) we first use (3) and (12) to obtain for all $x \in X$ and $y = Ax$

$$[\bar{T}(x) - \bar{T}[p(x)]]'[x - p(x)] = [T(Ax) - T[Ap(x)]]'[Ax - Ap(x)]$$
$$= [T(y) - T(y^*)]'(y - y^*) \geq \lambda |y - y^*|^2. \quad (25)$$

We now consider next the range $R(A)$ and the nullspace $N(A)$ of A, and the orthogonal complement

$$N(A)^+ = \{z \in \mathbf{R}^n \mid z'x = 0, \quad \forall x \in N(A)\}.$$

When viewed as a mapping from $N(A)^+$ to $R(A)$, A is one-to-one and onto, and since we are dealing with finite dimensional spaces we have that there exists a positive scalar $\beta(S)$ depending continuously on S such that

$$|w|^2 \geq \beta(S)\|z\|^2, \quad \forall w \in R(A), z \in N(A)^+, w = Az. \quad (26)$$

Since for all $x \in X$, $y = Ax$, and $\tilde{p}(x)$ defined by (20), we have

$$y - y^* = A[x - \tilde{p}(x)] \quad \text{and} \quad x - \tilde{p}(x) \in N(A)^+$$

we obtain from (26)

$$|y - y^*|^2 \geq \beta(S)\|x - \tilde{p}(x)\|^2. \quad (27)$$

Using the Pythagorean theorem and (21) we have

$$\|x - p(x)\|^2 = \|x - \tilde{p}(x)\|^2 + \|p(x) - \tilde{p}(x)\|^2$$
$$\leq \left[1 + \frac{1}{\eta(S)}\right]\|x - \tilde{p}(x)\|^2. \quad (28)$$

By combining (25), (27) and (28) we obtain

$$[\bar{T}(x) - \bar{T}[p(x)]]'[x - p(x)] \geq \lambda\beta(S)\left[1 + \frac{1}{\eta(S)}\right]^{-1}\|x - p(x)\|^2$$

so (24a) holds with

$$q(S) = \lambda\beta(S)\left[1 + \frac{1}{\eta(S)}\right]^{-1}.$$

To show (24b) we first use (2) and (12) to write

$$\|S^{-1}[\bar{T}(x) - \bar{T}[p(x)]]\|^2 = [T(y) - T(y^*)]'AS^{-1}A'[T(y) - T(y^*)]$$
$$\leq \Lambda(S)|T(y) - T(y^*)|^2 \leq \Lambda(S)L^2|y - y^*|^2 \quad (29)$$

where $\Lambda(S)$ is the largest eigenvalue of $AS^{-1}A'$.

By the same argument used to derive (26) we can assert that there exists a positive scalar $\gamma(S)$ depending continuously on S such that

$$|w|^2 \leq \gamma(S)\|z\|^2, \quad \forall w \in R(A), z \in N(A)^+, \quad w = Az. \quad (30)$$

Therefore

$$|y - y^*|^2 \leq \gamma(S)\|x - \tilde{p}(x)\|^2 \leq \gamma(S)\|x - p(x)\|^2. \quad (31)$$

Combination of (29) and (31) yields (24b) for $r(S) = \Lambda(S)L^2\gamma(S)$. This completes the proof of part (a).

(b) Let $\bar{\alpha}$ be any positive scalar such that $\bar{\alpha} < (2q(S))/r(S)$. Then using (18) we have for all $\alpha \in (0, \bar{\alpha}]$.

$$\|x_{k+1} - p(x_{k+1})\|^2 \leq t(\alpha)\|x_k - p(x_k)\|^2 \tag{32}$$

where $t(\alpha) = 1 - 2\alpha q(S) + \alpha^2 r(S)$. In view of the fact $\alpha \leq \bar{\alpha} < (2q(S))/r(S)$ it is easily seen that $0 \leq t(\alpha) < 1$. We have from (32)

$$\|x_k - p(x_k)\|^2 \leq t(\alpha)^k \|x_0 - p(x_0)\|^2 \tag{33}$$

and using (23) and (24) we obtain

$$\|x_{k+1} - p(x_k)\| \leq t(\alpha)^{1/2}\|x_k - p(x_k)\|. \tag{34}$$

By the triangle inequality

$$\|x_{k+1} - x_k\| \leq \|x_{k+1} - p(x_k)\| + \|x_k - p(x_k)\|. \tag{35}$$

Combining (35) with (33) and (34) we obtain

$$\begin{aligned}\|x_{k+1} - x_k\| &\leq [t(\alpha)^{1/2} + 1]\|x_k - p(x_k)\| \\ &\leq [t(\alpha)^{1/2} + 1]t(\alpha)^{k/2}\|x_0 - p(x_0)\|.\end{aligned} \tag{36}$$

Let $\beta = t(\alpha)^{1/2}$ and $\bar{q} = [t(\alpha)^{1/2} + 1]\|x_0 - p(x_0)\|$. Then (36) can be written as

$$\|x_{k+1} - x_k\| \leq \bar{q}\beta^k.$$

For all $k \geq 0$, $m \geq 1$ we have

$$\begin{aligned}\|x_{k+m} - x_k\| &\leq \|x_{k+m} - x_{k+m-1}\| + \cdots + \|x_{k+1} - x_k\| \\ &\leq \bar{q}(\beta^{k+m-1} + \cdots + \beta^k) = \frac{\bar{q}\beta^k(1-\beta^m)}{1-\beta}.\end{aligned} \tag{37}$$

Since in view of $t(\alpha) < 1$ we have $\beta < 1$, it follows that $\{x_k\}$ is a Cauchy sequence and hence converges to a vector x^*. Since by (33) $\{x_k - p(x_k)\}$ converges to zero we must have $x^* = p(x^*)$ which implies $x^* \in X^*$. By taking limit in (37) as $m \to \infty$ we obtain for all $k = 0, 1, \ldots$

$$\|x_k - x^*\| \leq q\beta^k$$

where $q = \bar{q}(1-\beta)^{-1}$. This completes the proof.

It is quite remarkable that as shown in Proposition 1(b), the sequence $\{x_k\}$ converges to a single limit point even though the set of solutions X^* may contain an infinite number of points.

3. An algorithm for the traffic assignment problem

In this section we consider the following network flow problem. A network consisting of a set of nodes $\mathcal{N}$ and a set of directed links $\mathcal{L}$ is given, together with a set W of node pairs referred to as *origin–destination* (*OD*) *pairs*. For OD pair $w \in W$ there is a known demand $d_w > 0$ representing traffic entering the network at the origin and exiting at the destination. For each OD pair w, the demand d_w is to be distributed among a given collection P_w of simple directed paths joining w. We denote by x_p the flow carried by path p. Thus the set of feasible path flow vectors $x = \{x_p \mid p \in P_w, w \in W\}$ is given by

$$X = \left\{x \,\middle|\, \sum_{p \in P_w} x_p = d_w, x_p \geq 0, \forall\, p \in P_w, w \in W\right\}. \tag{38}$$

Each collection of path flows $x \in X$ defines a collection of link flows y_{ij}, $(i, j) \in \mathcal{L}$ by means of the equation

$$y_{ij} = \sum_{w \in W} \sum_{p \in P_w} \delta_p(i, j) x_p, \quad \forall (i, j) \in \mathcal{L} \tag{39}$$

where $\delta_p(i, j) = 1$ if path p contains link (i, j) and $\delta_p(i, j) = 0$ otherwise. The vector of link flows $y = \{y_{ij} \mid (i, j) \in \mathcal{L}\}$ corresponding to $x \in X$ can be written as $y = Ax$ where A is the arc-chain matrix defined by (39). The set of feasible link flows is thus

$$Y = AX = \{y \mid y = Ax, x \in X\}. \tag{40}$$

We assume that for each link $(i, j) \in \mathcal{L}$ there is given a function $T_{ij}: Y \to \mathbf{R}$ such that $T_{ij}(y) > 0$ for all $y \in Y$. The value of $T_{ij}(y)$ represents a measure of delay in traversing link (i, j) when the set of link flows is y (travel time in transportation networks [1, 8], marginal delay in communication networks [4, 13]. The vector with components $T_{ij}(y)$ is denoted $T(y)$. We assume that $T(y)$ is Lipschitz continuous and strongly monotone. This is a reasonable assumption for transportation networks, as well as for communication networks.

For each $x \in X$ and corresponding $y = Ax$ the vector $T(y)$ defines for each $w \in W$ and $p \in P_w$ a length

$$\bar{T}_p(x) = \sum_{(i, j) \in \mathcal{L}} \delta_p(i, j) T_{ij}(y) \tag{41}$$

which may be viewed as the total travel time of path p. The problem is to find $x^* \in X$ such that for all $\bar{p} \in P_w$ and $w \in W$

$$\bar{T}_{\bar{p}}(x^*) = \min_{p \in P_w} \bar{T}_p(x^*), \quad \text{if } x^*_{\bar{p}} > 0. \tag{42}$$

This problem is based on the user-optimization principle which asserts that a traffic network equilibrium is established when no user may decrease his travel time by making a unilateral decision to change his route. If we denote by $\bar{T}(x)$ the vector of lengths $\bar{T}_p(x)$, $p \in P_w$, $w \in W$ then it is easy to see [1, 8] using (39) and (41) that

$$\bar{T}(x) = A'T(Ax) \tag{43}$$

and that the problem defined earlier via (42) is equivalent to finding a solution $x^* \in X$ of the variational inequality

$$(x - x^*)'\bar{T}(x^*) \geq 0, \quad \forall\, x \in X, \tag{44}$$

which is of the form (1).

Let $W = \{w_1, w_2, \ldots, w_M\}$ and consider the algorithm of the previous section with a matrix S which is block diagonal of the form

$$S = \begin{bmatrix} S_1 & & 0 \\ & S_2 & \\ 0 & & S_M \end{bmatrix}.$$

Here each matrix S_i corresponds to the OD pair w_i. We assume that each matrix S_i is positive definite symmetric. The projection iteration (4) can be implemented by finding x_{k+1} solving the quadratic program (16). In view of the block diagonal form of S and the decomposable nature of the constraint set X (cf. (38)) the quadratic program can be decomposed into a collection of smaller quadratic programs—one per OD pair $w \in W$. The form of these programs for the case where each matrix S_i is diagonal with elements s_p, $p \in P_{w_i}$ along the diagonal is

$$\begin{aligned} &\text{minimize} \quad \sum_{p \in P_{w_i}} \left\{ (x_p - x_p^k)\bar{T}_p(x_k) + \frac{s_p}{2\alpha}(x_p - x_p^k)^2 \right\}, \\ &\text{subject to} \quad \sum_{p \in P_{w_i}} x_p = d_{w_i}, \\ &\qquad\qquad\quad x_p \geq 0, \quad \forall\, p \in P_{w_i} \end{aligned} \tag{45}$$

where x_p^k, $p \in P_{w_i}$, $i = 1, \ldots, M$ are the components of the vector x_k. Problem (45) involves a single equality constraint and can be solved very easily—essentially in closed form [4, 12]. The convergence and linear rate of convergence results of Proposition 1 apply to this algorithm.

The preceding algorithm is satisfactory if each set P_{w_i} contains relatively few paths. In some problems however the number of paths in P_{w_i} can be very large (for example P_{w_i} may contain all simple paths joining w_i). In this case it is preferable to start with a subset of each P_{w_i} and augment this subset as necessary as suggested by Aashtiani [1]. The corresponding algorithm is as follows:

We begin with a subset $P_{w_i}^0 \subset P_{w_i}$ for each $w_i \in W$ and a vector of initial path flows x_0 such that for all $w_i \in W$

$$x_p^0 = 0, \quad \text{if } p \notin P_{w_i}^0.$$

At the kth iteration we have for each OD pair $w_i \in W$ a subset of paths $P^k_{w_i} \subset P_{w_i}$ and a corresponding vector of path flows x_k satisfying

$$x^k_p = 0, \quad \text{if } p \notin P^k_{w_i}.$$

We compute, for each $w_i \in W$, a shortest path $p^k_{w_i} \in P_{w_i}$ using $T_{ij}(Ax_k)$ as length for each link (i, j). We set

$$P^{k+1}_{w_i} = P^k_{w_i} \cup \{p^k_{w_i}\}.$$

(Note here that $p^k_{w_i}$ may already belong to $P^k_{w_i}$). We then solve the quadratic programming problem

$$\begin{aligned} &\text{minimize} \quad \sum_{p \in P^{k+1}_{w_i}} \left\{(x_p - x^k_p)\bar{T}_p(x_k) + \frac{S_p}{2\alpha}(x_p - x^k_p)^2\right\} \\ &\text{subject to} \quad \sum_{p \in P^{k+1}_{w_i}} x_p = d_{w_i}, \\ &\qquad\qquad\quad x_p \geq 0, \quad \forall\, p \in P^{k+1}_{w_i}. \end{aligned} \tag{46}$$

If $\bar{x}^k_p$, $p \in P^{k+1}_{w_i}$ is the solution of this problem we set

$$x^{k+1}_p = \begin{cases} \bar{x}^k_p, & \text{if } p \in P^{k+1}_{w_i}, \\ 0, & \text{if } p \notin P^{k+1}_{w_i}. \end{cases} \tag{47}$$

Note that the quadratic programming problem (46) is the same as (45) except for the fact that it involves only paths in the subset $P^{k+1}_{w_i}$. The subset is possibly augmented at each iteration k to include the current shortest path $p^k_{w_i}$. The expectation here is that, while P_{w_i} may contain a very large number of paths, the actual number of paths generated and included in the set $P^k_{w_i}$ remains small as k increases. This expectation has been supported by computational experiments [1]. The convergence and rate of convergence properties of the algorithm (46), (47) are identical with those of the earlier algorithm based on (45) as the reader can easily verify. The key idea is based on the fact that for each i, P_{w_i} contains a finite number of paths and $P^k_{w_i}$ grows monotonically so that the sequence $\{P^k_{w_i}\}$ converges to some subset of paths $\bar{P}_{w_i} \subset P_{w_i}$. By applying Proposition 1 it follows that the algorithm converges to a solution of the problem obtained when P_{w_i} is replaced by $\bar{P}_{w_i}$. Because $P^k_{w_i}$ is augmented at each iteration with the current shortest path it is a simple matter to conclude that the solution x^* obtained satisfies the required minimum travel time condition (42).

The results and algorithms of this section can be strengthened considerably in the case where T is in addition the gradient of convex function F in which case the problem is equivalent to the convex programming problem of minimizing $F(y)$ subject to $y \in Y$ or minimizing $F(Ax)$ subject to $x \in X$. Under these circumstances there are convergence results [4, 5, 12] relating to projection type algorithms which allow for a variable matrix S and for coordinate descent type iterations whereby each iteration is performed with respect to a single OD pair (or a small group of OD pairs) and all OD pairs are taken up in sequence. In fact for such algorithms it seems that it is easier to select an appropriate value for the stepsize α. Aashtiani's computational experience [1] suggests that such algorithms also work well in many cases where T is not a gradient mapping. We have been unable however to obtain a general convergence result for coordinate descent versions of the projection algorithm. The possibility of changing the matrix S from one iteration to the next is considered in the next section.

4. An algorithm with variable projection metric

A drawback of the algorithms of Section 2 and 3 is that the matrix S is restricted to be the same at each iteration. Computational experience with optimization problems as well as network flow problems [1, 6] suggests that, if T is differentiable, better results can be obtained if the matrix S is varied from one iteration to the next and is made suitably dependent on first derivatives of the mapping T in a manner which approximates Newton's method. We have not been able to show the result of Proposition 1(b) for algorithms in which the matrix S may change arbitrarily. On the other hand it is possible to construct an algorithmic scheme that allows for a variable matrix S but at the same time incorporates a mechanism that safeguards against divergence. The main idea in this scheme is *to allow a change in the matrix S only when the algorithm makes satisfactory progress towards convergence.* The algorithm is as follows:

A set $\mathcal{S}$ of positive definite symmetric matrices is given. It is assumed that all eigenvalues of all matrices $S \in \mathcal{S}$ lie in some compact interval of positive real numbers, i.e., there exist $m_1, m_2 > 0$ such that $m_1|z|^2 \leq z'Sz \leq m_2|z|^2$, for all $z \in \mathbf{R}^n$, $S \in \mathcal{S}$. We consider the algorithm

$$x_{k+1} = P_X^{S_k}[x_k - \alpha S_k^{-1}\bar{T}(x_k)], \quad x_0 \in X \tag{48}$$

where $S_k \in \mathcal{S}$ for all $k = 0, 1, \ldots$. The stepsize α is such that

$$h \overset{\Delta}{=} \max_{S \in \mathcal{S}}[1 - 2\alpha q(S) + \alpha^2 r(S)] < 1, \tag{49}$$

where $q(S)$ and $r(S)$ are as in Proposition 1(a) [c.f. (18)]. The maximum in (49) is attained by continuity of $q(\cdot)$ and $r(\cdot)$. It is clear that there exists $\bar{\alpha} > 0$ such that (49) is satisfied for all $\alpha \in (0, \bar{\alpha}]$. Given x_{k+1} the matrix S_{k+1} is either chosen arbitrarily from $\mathcal{S}$ or else $S_{k+1} = S_k$ depending on whether the quantity

$$w_k = (x_{k+1} - x_k)'S_k(x_{k+1} - x_k) \tag{50}$$

has decreased or not by a certain factor over the last time the matrix S was changed. More specifically a scalar $\bar{\beta} \in (0, 1)$ (typically close to unity) is chosen, and at each iteration k a scalar $\bar{w}_{k+1}$ is computed according to

$$\bar{w}_{k+1} = \begin{cases} \beta w_k, & \text{if } w_k \leq \bar{w}_k, \\ \bar{w}_k, & \text{if } w_k > \bar{w}_k \end{cases} \tag{51}$$

where w_k is given by (50) and initially $\bar{w}_0 = \infty$. We select

$$S_{k+1} = S_k, \quad \text{if } \bar{w}_{k+1} = \bar{w}_k, \tag{52a}$$

$$S_{k+1} \in \mathcal{S}, \quad \text{if } \bar{w}_{k+1} < \bar{w}_k. \tag{52b}$$

Thus for each k, the scalar $\bar{w}_k$ represents a target value below which w_k must drop in order for a change in S to be allowed in the next iteration.

We first show that if $\{x_k\}$ is a sequence generated by the algorithm just described, then

$$\liminf_{k\to\infty} w_k = 0. \tag{53}$$

Indeed if $\liminf_{k\to\infty} w_k > 0$, then S_k must have been allowed to change only a finite number of times in which case it follows from Propostion 1 that $\{x_k\}$ converges to a solution x^*. As a result we have $w_k \to 0$ contradicting the earlier assertion.

Let us denote by $\|\cdot\|_k$ the norm corresponding to S_k and by $p_k(z)$ the projection of a vector $z \in \mathbf{R}^n$ on X^* with respect to $\|\cdot\|_k$. We have by using the triangle inequality, (49), (50) and Proposition 1(a)

$$\begin{aligned} w_k^{1/2} &= \|x_{k+1} - x_k\|_k \\ &\geq \|x_k - p_k(x_{k+1})\|_k - \|x_{k+1} - p_k(x_{k+1})\|_k \\ &\geq \|x_k - p_k(x_k)\|_k - \sqrt{h}\|x_k - p_k(x_k)\|_k \\ &= (1-\sqrt{h})\|x_k - p_k(x_k)\|_k. \end{aligned} \tag{54}$$

Hence (53) implies $\liminf_{k\to\infty}\|x_k - p_k(x_k)\|_k = 0$ or equivalently (in view of the fact $\|z\|_k^2 \geq m_1|z|^2$, $m_1 > 0$)

$$\liminf_{k\to\infty} |x_k - p_k(x_k)| = 0. \tag{55}$$

This means that at least a subsequence of $\{x_k\}$ converges to the solution set X^*. We will show that in fact for some vector $x^* \in X^*$, and some scalars $q > 0$ and $\beta \in (0, 1)$ we have

$$|x_k - x^*| \leq q\beta^k, \quad \forall\, k = 0, 1, \ldots$$

i.e., $\{x_k\}$ converges to a solution x^* at a rate which is at least linear.

We have

$$\begin{aligned} w_k &= \|x_k - p_k(x_k) + p_k(x_k) - x_{k+1}\|_k^2 \\ &\leq \|x_k - p_k(x_k)\|_k^2 + \|p_k(x_k) - x_{k+1}\|_k^2 \\ &\quad + 2\|x_k - p_k(x_k)\|_k\|p_k(x_k) - x_{k+1}\|_k. \end{aligned} \tag{56}$$

Using (23), (24) and (49) we obtain

$$\|p_k(x_k) - x_{k+1}\|_k^2 \le h\|x_k - p_k(x_k)\|_k^2. \tag{57}$$

Combination of (56) and (57) yields

$$w_k \le (1+\sqrt{h})^2\|x_k - p_k(x_k)\|_k^2. \tag{58}$$

Also in view of (49) and Proposition 1 we have that there exists a scalar $d \ge 1$ such that for all k

$$\begin{aligned}\|x_{k+1} - p_{k+1}(x_{k+1})\|_{k+1}^2 &\le \|x_{k+1} - p_k(x_{k+1})\|_{k+1}^2\\ &\le d\|x_{k+1} - p_k(x_{k+1})\|_k^2\\ &\le dh\|x_k - p_k(x_k)\|_k^2\end{aligned}$$

and finally using (54)

$$\|x_{k+1} - p_{k+1}(x_{k+1})\|_{k+1}^2 \le \frac{dh}{(1-\sqrt{h})^2} w_k. \tag{59}$$

Let

$$\mathcal{K} = \{k \mid w_k \le \bar{w}_k\}. \tag{60}$$

By (51) and (52) we have

$$S_{k+1} = S_k, \quad \forall\, k \notin \mathcal{K}. \tag{61}$$

Also from (51) and (53) and the fact that if $w_{\bar{k}} = 0$, then $w_k = 0$ for all $k \ge \bar{k}$, it follows that $\mathcal{K}$ contains an infinite number of indices. Let k_1 and k_2 be two successive indices in $\mathcal{K}$ with $k_1 < k_2$. Then

$$w_{k_2} \le \bar{w}_{k_2} = \bar{\beta} w_{k_1} \tag{62}$$

while, if $k_2 - k_1 > 1$, we have

$$w_{k_1+m} > \bar{w}_{k_2} = \bar{\beta} w_{k_1}, \quad \forall\, m = 1, \dots, (k_2 - k_1 - 1). \tag{63}$$

We also have $S_{k_1+1} = \cdots = S_{k_2}$. In the case where $k_2 - k_1 > 1$, Proposition 1(a) together with (49) and (59) yields for all $m = 1, \dots, (k_2 - k_1 - 1)$

$$\begin{aligned}\|x_{k_1+m} - p_{k_1+m}(x_{k_1+m})\|_{k_1+1}^2 &\le h^{m-1}\|x_{k_1+1} - p_{k_1+1}(x_{k_1})\|_{k_1+1}^2\\ &\le \frac{dh^{m-1}}{(1-\sqrt{h})^2} w_{k_1}.\end{aligned} \tag{64}$$

Using (58) and (64) we have

$$w_{k_1+m} \le \frac{dh^{m-1}(1+\sqrt{h})^2}{(1-\sqrt{h})^2} w_{k_1}, \quad \forall\, m = 1, \dots, (k_2 - k_1 - 1). \tag{65}$$

Inequalities (63) and (65) yield

$$\bar{\beta} w_{k_1} < w_{k_2-1} \le \frac{dh^{k_2-k_1-2}(1+\sqrt{h})^2}{(1-\sqrt{h})^2} w_{k_1}.$$

It follows that if $k_2 - k_1 \geq 1$, then

$$k_2 - k_1 \leq 2 + \frac{\ln \dfrac{\bar{\beta}(1-\sqrt{\bar{h}})^2}{d(1+\sqrt{\bar{h}})^2}}{\ln h}, \tag{66}$$

so if $\bar{m}$ is any positive integer such that

$$2 + \frac{\ln \dfrac{\bar{\beta}(1-\sqrt{\bar{h}})^2}{d(1+\sqrt{\bar{h}})^2}}{\ln h} \leq \bar{m} \tag{67}$$

we have

$$k_2 - k_1 \leq \bar{m} \tag{68}$$

for any two successive indices k_1, k_2 in $\mathcal{K}$. It follows using (62) that

$$w_k \leq w_0(\bar{\beta}^{1/\bar{m}})^k, \quad \forall\, k \in \mathcal{K}. \tag{69}$$

Using (65), (68) and (69) we also obtain

$$w_k \leq \frac{d(1+\sqrt{\bar{h}})^2 w_0}{(1-\sqrt{\bar{h}})^2}(\bar{\beta}^{1/\bar{m}})^{k-\bar{m}}, \quad \forall\, k \notin \mathcal{K}. \tag{70}$$

Combining (69) and (70) we have for some scalar $\bar{q} > 0$

$$w_k \leq \bar{q}(\bar{\beta}^{1/\bar{m}})^k, \quad \forall k = 0, 1, \ldots \tag{71}$$

Since

$$m_1|x_{k+1} - x_k|^2 \leq \|x_{k+1} - x_k\|_k^2 = w_k$$

relation (71) yields

$$|x_{k+1} - x_k|^2 \leq \frac{\bar{q}}{m_1}(\bar{\beta}^{1/\bar{m}})^k, \quad \forall\, k = 0, 1, \ldots$$

Since $\bar{\beta} < 1$, it follows in exactly the same manner as in the proof of Proposition 1(b) that $\{x_k\}$ is a Cauchy sequence which converges to a vector $x^* \in X^*$. Furthermore for some $q > 0$ and $\beta \in (0, 1)$ we have

$$|x_k - x^*| \leq q\beta^k, \quad k = 0, 1, \ldots \tag{72}$$

We have thus proved the following proposition.

Proposition 2. *There exists $\bar{\alpha} > 0$ such that if $\alpha \in (0, \bar{\alpha}]$, a sequence $\{x_k\}$ generated by iteration (48) with $\{S_k\}$ selected according to (50)–(52) converges to a solution x^* of the variational inequality (1). Furthermore there exist scalars $q > 0$ and $\beta \in (0, 1)$ such that*

$$|x_k - x^*| \leq q\beta^k, \quad \forall\, k = 0, 1, \ldots$$

5. A computational example

In this section we report computational results for a traffic assignment problem. The corresponding network is shown in Fig. 2, and may be viewed as a model of a circular highway. There are five origins and destinations numbered 1, 2, 3, 4, 5 and connected through the highway via entrance and exit ramps. We consider the five OD pairs (1, 4), (2, 5), (3, 1), (4, 2), (5, 3). Each OD pair has two paths associated with it—the clockwise and counterclockwise paths on the corresponding circle. The expressions for the travel time on each link are shown in Fig. 2 where the function g is given by $g(x) = 1 + x + x^2$. Different values of the nonnegative scalar γ represent different degrees of dependence of the travel times of some links on the flows of other links. The problem is equivalent to an optimization problem if and only if there is no such dependence ($\gamma = 0$).

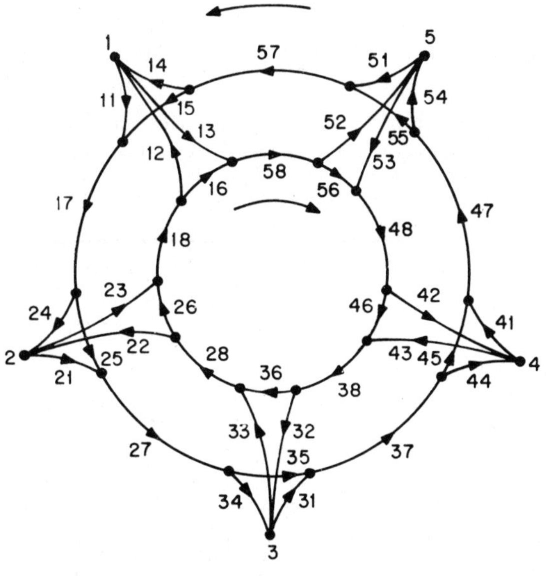

Fig. 2.

Types of links:
(1) Highway links: 17, 27, 37, 47, 57, 18, 28, 38, 48, 58.
(2) Exit ramps: 14, 24, 34, 44, 54, 12, 22, 32, 42, 52.
(3) Entrance ramps: 11, 21, 31, 41, 51, 13, 23, 33, 43, 53.
(4) Bypass links: 15, 25, 35, 45, 55, 16, 26, 36, 46, 56.

Delay on links [where g is defined by $g(x) = 1 + x + x^2$]:
(1) Delay on highway link k: $10 \cdot g$[Flow on k] $+ 2 \cdot \gamma \cdot g$[Flow on exit ramp from k].
(2) Delay on exit ramp k: g[Flow on k].
(3) Delay on entrance ramp k: g[Flow on k] $+ \gamma \cdot g$[Flow on bypass link merging with k].
(4) Delay on bypass link k: g[Flow on k].

Remark: Flow is not allowed to use exit ramp not leading to its destination.
OD Pairs: (1, 4), (2, 5), (3, 1), (4, 2), (5, 3).

Tables 1 and 2 list representative computational results for two demand patterns, three values of γ, and fifteen iterations of two different algorithms labeled 'all-at-once' and 'one-at-a-time' and described below. The number shown in the tables for each iteration k is the following normalized measure of convergence

$$\sum_{\text{all OD pairs } w} \frac{\Delta x_{w,k}}{d_w} \frac{\Delta \bar{T}_{w,k}}{\bar{T}_{\min, w, k}} \tag{73}$$

where d_w is the demand of the OD pair w, $\Delta x_{w,k}$ is the portion of the demand that does not lie on the shortest path of the OD pair w at the end of iteration k, $\Delta \bar{T}_{w,k}$ is the difference of the travel times of the longest and shortest paths and $\bar{T}_{\min, w, k}$ is the travel time of the shortest path. Clearly the expression (73) is zero if and only if the corresponding traffic assignment is optimal. The starting flow pattern in all runs was the worst possible whereby all the demand of each OD pair is routed on the counter-clockwise path. The results suggest that the algorithms yield near optimal flow patterns after very few iterations and subsequently continue their progress at a fairly satisfactory rate. This type of convergence behavior is consistent with the one observed for related algorithms tested in [6].

The 'all-at-once' algorithm is the one of the previous section [c.f. (48), (50)–(52)] with the projection matrix S_k being diagonal [c.f. (45)]. For each

Table 1
Demands $d(1,4) = 0.1$; $d(2,5) = 0.2$; $d(3,1) = 0.3$; $d(4,2) = 0.4$; $d(5,3) = 0.5$

	All-at-once, $\alpha = 0.8$, $\bar{\beta} = 0.99$			One-at-a-time, $\alpha = 1$, $\bar{\beta} = 0.99$		
K	$\gamma = 0$	$\gamma = 0.5$	$\gamma = 4$	$\gamma = 0$	$\gamma = 0.5$	$\gamma = 4$
0	0.14417×10^{2}	0.14793×10^{2}	0.17426×10^{2}	0.14417×10^{2}	0.14793×10^{2}	0.17426×10^{2}
1	0.14897×10^{1}	0.15079×10^{1}	0.16436×10^{1}	0.43831×10^{0}	0.46175×10^{0}	0.36140×10^{0}
2	0.39463×10^{0}	0.36291×10^{0}	0.23633×10^{0}	0.73026×10^{-1}	0.80765×10^{-1}	0.13319×10^{0}
3	0.35901×10^{0}	0.29642×10^{0}	0.12340×10^{0}	0.28867×10^{-1}	0.37109×10^{-1}	0.39248×10^{-1}
4	0.55230×10^{-1}	0.41860×10^{-1}	0.11996×10^{-1}	0.11671×10^{-1}	0.17501×10^{-1}	0.15697×10^{-1}
5	0.80434×10^{-1}	0.52264×10^{-1}	0.55236×10^{-2}	0.50979×10^{-2}	0.86871×10^{-2}	0.10786×10^{-1}
6	0.11485×10^{-1}	0.70972×10^{-2}	0.14528×10^{-2}	0.23552×10^{-2}	0.45007×10^{-2}	0.76687×10^{-2}
7	0.19034×10^{-1}	0.93712×10^{-2}	0.71098×10^{-3}	0.11224×10^{-2}	0.23972×10^{-2}	0.55389×10^{-2}
8	0.26034×10^{-2}	0.13097×10^{-2}	0.34892×10^{-3}	0.54303×10^{-3}	0.12962×10^{-2}	0.40331×10^{-2}
9	0.43683×10^{-2}	0.21587×10^{-2}	0.18026×10^{-3}	0.26555×10^{-3}	0.70754×10^{-3}	0.29498×10^{-2}
10	0.55167×10^{-3}	0.11845×10^{-2}	0.99838×10^{-4}	0.12989×10^{-3}	0.38809×10^{-3}	0.21641×10^{-2}
11	0.10228×10^{-2}	0.11548×10^{-2}	0.80656×10^{-4}	0.64334×10^{-4}	0.21363×10^{-3}	0.15911×10^{-2}
12	0.44825×10^{-3}	0.80420×10^{-3}	0.68090×10^{-4}	0.31281×10^{-4}	0.11782×10^{-3}	0.11698×10^{-2}
13	0.51164×10^{-3}	0.67664×10^{-3}	0.59269×10^{-4}	0.15289×10^{-4}	0.64249×10^{-4}	0.86171×10^{-3}
14	0.30760×10^{-3}	0.51037×10^{-3}	0.51284×10^{-4}	0.77946×10^{-5}	0.35938×10^{-4}	0.63501×10^{-3}
15	0.27834×10^{-3}	0.41039×10^{-3}	0.44031×10^{-4}	0.41734×10^{-5}	0.19540×10^{-4}	0.46808×10^{-3}

Table 2
Demands $d(1,4)=1$, $d(2,5)=8$, $d(3,1)=1$, $d(4,2)=8$, $d(5,3)=1$

	All-at-once, $\alpha=0.8$, $\bar{\beta}=0.99$			One-at-a-time, $\alpha=1$, $\bar{\beta}=0.99$		
K	$\gamma=0.0$	$\gamma=0.5$	$\gamma=4$	$\gamma=0$	$\gamma=0.5$	$\gamma=4$
0	0.10203×10^{4}	0.10478×10^{4}	0.12404×10^{4}	0.10203×10^{4}	0.10478×10^{4}	0.12405×10^{4}
1	0.19446×10^{1}	0.15853×10^{1}	0.33203×10^{0}	0.28891×10^{1}	0.24262×10^{1}	0.86532×10^{0}
2	0.83731×10^{0}	0.57695×10^{0}	0.27719×10^{-1}	0.65950×10^{0}	0.49153×10^{0}	0.59564×10^{-1}
3	0.12902×10^{1}	0.90467×10^{0}	0.37155×10^{-1}	0.78641×10^{-1}	0.56482×10^{-1}	0.13636×10^{-1}
4	0.45269×10^{0}	0.24701×10^{0}	0.24096×10^{-1}	0.63059×10^{-2}	0.70762×10^{-2}	0.56463×10^{-2}
5	0.83315×10^{0}	0.42439×10^{0}	0.17847×10^{-1}	0.70066×10^{-3}	0.27832×10^{-2}	0.29389×10^{-2}
6	0.23621×10^{0}	0.79849×10^{-1}	0.12985×10^{-1}	0.10199×10^{-3}	0.12097×10^{-2}	0.16215×10^{-2}
7	0.50649×10^{0}	0.11570×10^{0}	0.95892×10^{-2}	0.79205×10^{-4}	0.45572×10^{-3}	0.90603×10^{-3}
8	0.11333×10^{0}	0.10000×10^{-1}	0.71098×10^{-2}	0.14878×10^{-4}	0.16733×10^{-3}	0.50823×10^{-3}
9	0.26874×10^{0}	0.78683×10^{-2}	0.52748×10^{-2}	0.94583×10^{-4}	0.61370×10^{-4}	0.28569×10^{-3}
10	0.43230×10^{-1}	0.10397×10^{-2}	0.39184×10^{-2}	0.65018×10^{-4}	0.22825×10^{-4}	0.16119×10^{-3}
11	0.11434×10^{0}	0.13673×10^{-2}	0.29147×10^{-2}	0.39872×10^{-4}	0.90891×10^{-5}	0.90105×10^{-4}
12	0.12722×10^{-1}	0.68513×10^{-3}	0.21694×10^{-2}	0.24651×10^{-4}	0.33388×10^{-5}	0.51215×10^{-4}
13	0.43756×10^{-1}	0.50364×10^{-3}	0.16170×10^{-2}	0.15648×10^{-4}	0.10026×10^{-5}	0.27883×10^{-4}
14	0.45550×10^{-2}	0.32649×10^{-3}	0.12047×10^{-2}	0.92416×10^{-5}	0.40281×10^{-6}	0.15870×10^{-4}
15	0.20089×10^{-1}	0.22516×10^{-3}	0.89921×10^{-3}	0.68895×10^{-5}	0.47333×10^{-6}	0.89927×10^{-5}

iteration k for which S_k was allowed to change [c.f. (50)–(52)] the diagonal element of S_k corresponding to any one path was taken equal to the sum of the first derivatives of the travel times of links on that path evaluated at the kth flow x_k. This corresponds to a diagonal approximation of Newton's method (c.f. [4, 5]). As a result stepsizes near unity typically give satisfactory convergence behavior. In all runs we used a stepsize $\alpha=0.8$ which is probably a bit on the high side. The scalar $\bar{\beta}$ used in the test for allowing the matrix S_k to change [c.f. (51)] was taken to be 0.99 in all runs. One of the most interesting observations from our experimentation was that this test was passed at every iteration and so *the matrix S_k was changed at every iteration.*

The 'one-at-a-time' algorithm is similar to the 'all-at-once' algorithm described above. The only difference is that the projection is carried out with respect to a single OD pair with flows corresponding to the other OD pairs being kept fixed, and all OD pairs are taken up in sequence. An iteration consists of a cycle of five projection subiterations (one per OD pair). Each subiteration is, of course, followed by reevaluation of the travel time of each link. Algorithms of this type resemble coordinate descent methods for unconstrained optimization and have been suggested in the context of network flows in [1, 4, 5, 12]. The stepsize α was taken to be unity in all runs. Also $\bar{\beta}$ was chosen to be 0.99 and again it turned out that the matrix S_k was allowed to change in every iteration. It is interesting to note that for this stepsize and this particular example the algorithm tested is equivalent to an algorithm in the class proposed by Aashtiani [1]. Tables 1 and 2 indicate a better performance for the 'one-at-a-time' algorithm. However, there is no convergence proof available for this algorithm at present except in the case where the corresponding variational inequality is equivalent to a convex optimization problem.

Appendix. Proof of Lemma 1

We have by the Pythagorean theorem

$$\|x - \tilde{p}(x)\|^2 + \|\tilde{p}(x) - p(x)\|^2 = \|x - p(x)\|^2 \le \|x - p[\tilde{p}(x)]\|^2$$
$$= \|x - \tilde{p}(x)\|^2 + \|\tilde{p}(x) - p[\tilde{p}(x)]\|^2$$

Hence

$$\|\tilde{p}(x) - p(x)\|^2 \le \|\tilde{p}(x) - p[\tilde{p}(x)]\|^2. \tag{A.1}$$

Since $p[\tilde{p}(x)]$ by definition is the unique solution of the problem of minimizing $\|\tilde{p}(x) - z\|^2$ over $x \in X^*$, and $p(x) \in X^*$, it follows that equality holds in (A.1) and $p(x) = p[\tilde{p}(x)]$.

Let $\tilde{p}_1(x)$ and $p_1(x)$ be the projections of x on $\tilde{X}^*$ and X^* respectively relative to the standard norm $|\cdot|$, i.e., $\tilde{p}_1(x) = \tilde{p}(x)$, $p_1(x) = p(x)$ for S equal to the identity. In order to show the existence of a continuous $\eta(S)$ such that (21) holds it will suffice to show the existence of a scalar $\eta_1 > 0$ such that for all $x \in X$

$$|x - \tilde{p}_1(x)|^2 \ge \eta_1 |p_1(x) - \tilde{p}_1(x)|^2. \tag{A.2}$$

In order to see this let $\Lambda(S)$ and $\lambda(S)$ be the largest and smallest eigenvalue of S. We have for all x, $z \in \mathbf{R}^n$

$$\lambda(S)|x - z|^2 \le \|x - z\|^2 \le \Lambda(S)|x - z|^2.$$

It follows that

$$\frac{1}{\lambda(S)}\|x - \tilde{p}(x)\|^2 \ge |x - \tilde{p}_1(x)|^2 \tag{A.3}$$

$$\frac{1}{\Lambda(S)}\|x - p(x)\|^2 \le |x - p_1(x)|^2. \tag{A.4}$$

By the Pythagorean theorem we have

$$|p_1(x) - \tilde{p}_1(x)|^2 = |x - p_1(x)|^2 - |x - \tilde{p}_1(x)|^2$$

so (A.2) can be rewritten as

$$|x - \tilde{p}_1(x)|^2 \ge \frac{\eta_1}{1 + \eta_1}|x - p_1(x)|^2. \tag{A.5}$$

From (A.3), (A.4) and (A.5) we obtain

$$\|x - \tilde{p}(x)\|^2 \geq \gamma(S)\|x - p(x)\|^2 \tag{A.6}$$

where

$$\gamma(S) = \frac{\eta_1}{1+\eta_1} \frac{\lambda(S)}{\Lambda(S)}.$$

By the Pythagorean theorem we have

$$\|x - p(x)\|^2 = \|x - \tilde{p}(x)\|^2 + \|p(x) - \tilde{p}(x)\|^2$$

and by using this equation in (A.6) we obtain the desired relation

$$\|x - \tilde{p}(x)\|^2 \geq \eta(S)\|p(x) - \tilde{p}(x)\|^2$$

with

$$\eta(S) = \frac{\gamma(S)}{1 - \gamma(S)}.$$

We now show existence of an $\eta_1 > 0$ such that (A.2) holds for all $x \in X$. By the preceding analysis this is sufficient to prove the lemma.

For each $x \in X$ consider the tangent cone C_x of X at $p(x)$, i.e., the set

$$C_x = \{z \mid \text{there exists } \alpha > 0 \text{ such that } [p(x) + \alpha z] \in X\}. \tag{A.7}$$

Let $\mathscr{C}$ be the collection

$$\mathscr{C} = \{C_x \mid x \in X\}. \tag{A.8}$$

It is easily seen that $\mathscr{C}$ *is a finite collection*, i.e., for some finite set $J \subset X$ we have

$$\mathscr{C} = \{C_j \mid j \in J\}. \tag{A.9}$$

Indeed since X is polyhedral it can be represented by definition [15, Section 19], in terms of a finite number of vectors $d_1, \ldots, d_m \in \mathbf{R}^n$ and scalars $b_1, \ldots, b_m$ as

$$X = \{x \mid d_i'x \leq b_i,\ i = 1, \ldots, m\}.$$

It is easily seen that

$$C_x = \{z \mid d_i'z \leq 0,\ \forall\, i = 1, \ldots, m \text{ such that } d_i'p(x) = b_i\}.$$

Clearly there is only a finite number of sets of the above form. In what follows $P_\Omega(z)$ denotes the projection of a vector $z \in \mathbf{R}^n$ on a closed convex set $\Omega \subset \mathbf{R}^n$ with respect to the standard norm $|\cdot|$. The essence of the proof of Lemma 1 is contained in the following lemma.

Lemma A.1. *For $j \in J$ let*

$$M_j^* = C_j \cap N(A), \tag{A.10}$$

$$Z_j = \{z \mid z \in C_j, P_{M_j^*}(z) = 0\}, \tag{A.11}$$

where $N(A)$ is the nullspace of A. Then for each $j \in J$ there exists a scalar $\eta_j > 0$ such that

$$|z - P_{N(A)}(z)|^2 \geq \eta_j |z|^2, \quad \forall z \in Z_j \tag{A.12}$$

Proof. Assume the contrary, i.e., that there exist $j \in J$ and sequences $\{z_k\} \subset Z_j$, $\{\eta_j^k\} \subset R$ such that

$$|z_k - P_{N(A)}(z_k)|^2 < \eta_j^k |z_k|, \quad \eta_j^k \to 0. \tag{A.13}$$

We then have $z_k \neq 0$, $\forall k$, and since both M_j^* and Z_j are clearly cones with vertex at the origin we can assume that $|z_k| = 1$, $\forall k$. Let $\bar{z}$ be a limit point of $\{z_k\}$. We have $|\bar{z}| = 1$ and by taking limit in (A.13) we obtain $\bar{z} = P_{N(A)}(\bar{z})$ i.e., $\bar{z} \in N(A)$. Since C_j is closed we also have $\bar{z} \in C_j$ and hence $\bar{z} \in M_j^*$. It follows that

$$P_{M_j^*}(\bar{z}) = \bar{z}. \tag{A.14}$$

On the other hand since $z_k \in Z_j$ we have $P_{M_j^*}(z_k) = 0$, $\forall k$ which implies that $P_{M_j^*}(\bar{z}) = 0$. Since $|\bar{z}| = 1$, this contradicts (A.14).

We now show that the desired relation (A.2) holds with

$$\eta_1 = \min\{\eta_j \mid j \in J\} > 0. \tag{A.15}$$

Choose any $x \in X$ and let $j \in J$ be such that $C_x = C_j$. Let

$$z = x - p_1(x). \tag{A.16}$$

By a simple translation argument, the fact that $p_1(x)$ is the projection of x on $X \cap \{x \mid Ax = y^*\}$ implies that the projection of z on M_j^* is the origin so that $z \in Z_j$. A similar argument shows that

$$\tilde{p}_1(x) - p_1(x) = P_{N(A)}(z). \tag{A.17}$$

Using (A.16) and (A.17) in (A.12) we obtain

$$|x - \tilde{p}_1(x)|^2 \geq \eta_j |x - p_1(x)|^2$$

and (A.2) follows from the definition (A.15) of η_1.

References

[1] H.Z. Aashtiani, "The multi-model assignment problem", Ph.D. Thesis, Sloan School of Management, Massachusetts Institute of Technology (May, 1979).

[2] A. Auslender, *Optimization. Méthodes numériques* (Mason, Paris, 1976).

[3] A.B. Bakushinskij and B.T. Poljak, "On the solution of variational inequalities", *Soviet Mathematics Doklady* 219 (1974) 1705–1710.

[4] D.P. Bertsekas, "Algorithms for nonlinear multicommodity network flow problems", in: A. Bensoussan and J.L. Lions, eds., *International Symposium on Systems Optimization and Analysis* (Springer, New York, 1979) pp. 210–224.

[5] D.P. Bertsekas, "A class of optimal routing algorithms for communication networks", Proceedings of 1980 ICCC, Atlanta, GA (1980) pp. 71–75.

[6] D.P. Bertsekas, E. Gafni and K.S. Vastola, "Validation of algorithms for routing of flow in networks", Proceedings of 1978 Conference on Decision and Control, San Diego, CA (1979) pp. 220–227.

[7] D.G. Cantor and M. Gerla, "Optimal routing in a packet switched computer network", *IEEE Transactions on Computers* C-23 (1974) 1062–1069.

[8] S. Dafermos, "Traffic equilibrium and variational inequalities", *Transportation Science* 14 (1980) 42–54.

[9] J.C. Dunn, "Global and asymptotic convergence rate estimates for a class of projected gradient processes", *SIAM Journal on Control and Optimization* 19 (1981) 368–400.

[10] J.C. Dunn, "Newton's method and the Goldstein step length rule for constrained minimization problems", *SIAM Journal on Control and Optimization* 18 (1980) 659–674.

[11] L. Fratta, M. Gerla, and L. Kleinrock, "The flow deviation method: An approach to store- and forward communication network design", *Networks* 3 (1973) 97–133.

[12] E.M. Gafni, "Convergence of a routing algorithm", Laboratory for Information and Decision Systems Report 907, Massachussetts Institute of Technology, Cambridge, MA (May 1979).

[13] R.G. Gallager, "A minimum delay routing algorithm using distributed computation", *IEEE Transactions on Communication* COM-25 (1977) 73–85.

[14] E.S. Levitin and B.T. Poljak, "Constrained minimization methods", *U.S.S.R. Computational Mathematics and Mathematical Physics* 6 (1966) 1–50.

[15] R.T. Rockafellar, *Convex analysis* (Princeton University Press, Princeton, 1970).

[16] M. Sibony, "Méthodes itératives pour les équations et inéquations aux dérivées partielles nonlinéares de type monotone", *Calcolo* 7 (1970) 65–183.

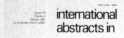

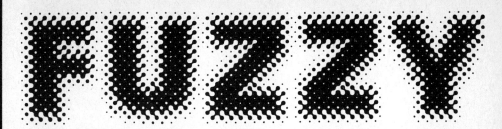